ECOLOGY AND LIFE

**Accepting Our
Environmental Responsibility**

ISSUES OF CHRISTIAN CONSCIENCE

Nuclear Arms: Two Views on World Peace
Myron S. Augsburger and Dean C. Curry

Ecology and Life: Accepting Our Environmental Responsibility
Wesley Granberg-Michaelson

ECOLOGY AND LIFE

Accepting Our Environmental Responsibility

Wesley Granberg-Michaelson

General Editor
Vernon Grounds

ISSUES OF CHRISTIAN CONSCIENCE

WORD BOOKS
PUBLISHER
WACO, TEXAS

A DIVISION OF
WORD, INCORPORATED

ECOLOGY AND LIFE
Issues of Christian Conscience

Unless otherwise noted, all Scripture quotations are from the Revised Standard Version of the Bible, copyrighted 1946, 1952, 1971 by the Division of Christian Education of the National Council of Churches of Christ in the U.S.A., and are used by permission. All rights reserved. Quotations identified as NEB are from The New English Bible, © The Delegates of The Oxford University Press and the Syndics of The Cambridge University Press, 1961, 1970.

An effort has been made to locate sources and obtain permission where necessary for quotations used in this book. In the event of any omission, a modification and acknowledgment will gladly be incorporated in future editions.

The Epilogue, "Fly-fishing As a Spiritual Discipline," appeared in *Perspectives*, September 1986. Certain portions of chapters 7 and 8 also appeared, in earlier versions, in *Perspectives*.

"Nature, Humanity, and Biblical Theology" appeared in edited form in *The Predicament of the Prosperous*, by Bruce C. Birch and Larry L. Rasmussen, © 1978 The Westminster Press. Used by permission. "Theocentrism: The Cornerstone of Christian Ecology" by Vincent Rossi first appeared in the Fall 1985 issue of *Epiphany: A Journal of Faith and Insight* (P.O. Box 14727, San Francisco, CA 94117) and is reprinted by permission. "'All Creation Groans': Theology/Ecology in St. Paul" by James A. Rimbach first appeared in the *Asia Journal of Theology* 1:2 (1987), pp. 379–91, and is used by permission. "God's Joyous Valuing of Nature" is reprinted with permission from *Brother Earth* by H. Paul Santmire, © 1970, published by Thomas Nelson. "The Historical Roots of Our Ecologic Crisis," by Lynn White, Jr. is reprinted from *Science*, Vol. 155, pp. 1203–07, March 10, 1967. © 1967 by the American Association for the Advancement of Science.

Library of Congress Cataloging-in-Publication Data

Granberg-Michaelson, Wesley.
    Ecology and life : accepting our environmental responsibility /
Wesley Granberg-Michaelson.
       p.   cm. — (Issues of Christian conscience)
    Bibliography: p.
    Includes index.
    ISBN 0-8499-0579-6
    1. Human ecology—Religious aspects—Christianity.
2. Evangelicalism. I. Title. II. Series.
BT695.5.G72 1988
261.8'362—dc19   88-10064
                                          CIP

Printed in the United States of America
89801239 RRD 987654321

To Dad and Mom,

for all they have given to me.

Contents

Foreword

In the New Testament, Christians are repeatedly instructed to avoid evil and do good. For example, the apostle Peter in his first Letter exhorts his fellow believers to "turn from evil and do good" (1 Pet. 3:11), an admonition quoted directly from Psalm 34. So the entire Bible, both Old Testament and New, teaches that good is to be done, evil avoided.

But how can we determine what is good over against what is evil? Scripture, of course, gives some specific moral directives and lays down broad behavior-controlling principles like, "Love your neighbor as yourself." Yet in countless situations which we encounter in living out our discipleship, the Bible furnishes us with no explicit guidance.

Do we or do we not turn off the respirator when an aged family member is comatose and incurably ill? Do we or do we not engage in artificial insemination? Do we or do we not sanction remarriage after divorce? Do we or do we not advocate the use of atomic power for peaceful purposes—at the risk of lethally contaminating the environment? Do we or do we not favor legislation which mandates extensive and undeniably expensive welfare programs? Do we or do we not approve a vast buildup of armaments in our quest for world peace? Do we or do we not share the conviction that military strength is the only means of insuring national security and preventing global destruction— at least to a high degree of probability? Do we or do we not endorse a vigorous policy of aggressive anti-communism?

These are merely a few of the issues which we face today, issues to which Scripture does not speak unambiguously. Hence, as wisely and prayerfully as possible, with the help of insights gleaned from our spiritual traditions and communities, we struggle to apply those broad biblical guidelines to our moral problems. By the best logic of which we are capable, we attempt to deduce concrete decisions from the generalized moral imperatives we find in the Bible. And in many cases, that proves to be a difficult process indeed, one which results in the clash of radically opposing viewpoints among equally sincere, rational, and well-informed Christians.

God, however, holds each of us personally responsible for our choices. He expects us according to our abilities to read, think, discuss, learn, and pray. He knows the measure of our intelligence, the scope of our opportunities for ascertaining the truth, the circumstances that may inescapably warp our perceptions, the enervating factors that can cause cognitive sluggishness, and the indifference for which we are culpable. He alone knows whether we have availed ourselves of the information within our reach regarding some of the conclusions we draw and the actions we are existentially constrained to take. Only he, consequently, is able to judge us infallibly—but judge us he will for our errors, failures, and sins in this area of responsibility, as in all others.

Thus the purpose of the Issues of Christian Conscience series is to make available discussions by experts of some of the most important problems that we confront in the late twentieth century—and some of the most hotly debated.

Though all the authors will invariably be adherents and protagonists of definite viewpoints, they will nevertheless present their interpretations and convictions as objectively and persuasively as they are able. In short, this Word project is designed to be informative, not propagandistic.

This second volume in the series focuses attention on ecology, a crucial issue indeed. How can we stop the pollution of our planet, the deadly defilement of our global home, the short-sighted, self-destructive activities now being carried on by many members of the human crew on spaceship earth? My hope is that, as we read Wesley Granberg-Michaelson's plea for environmental responsibility, we will become concerned enough to modify our own lifestyles and at the same time work for those economic and political changes that are desperately imperative.

It is our intention to inform the Christian community not only of the importance of such issues, but of the responsibilities we face regarding such issues, as well. Then, as Christians opt for a certain behavioral pattern or political policy, their choices will hopefully be made with an enlightened conscientiousness.

VERNON GROUNDS
General Editor

1

Questions of a Christian

Randy, a young college student in our church, came into my office one afternoon with pressing questions on his heart. He had become a Christian through a Bible study at a ranch in Colorado three years earlier. Prior to that time, he had been an environmental science major in school and had been involved in the work of several environmental organizations. But since becoming a Christian, he had been shocked to discover, first of all, that many scholars and historians, as well as some environmental activists, blamed Christianity for the environmental crisis. Second, he was dismayed that fellow evangelicals simply didn't care about the environmental issue, and didn't sympathize with his concerns.

Randy's observations of nature were one of the influences that led him on his journey into Christian faith.

"When you look at mountains, columbines, and hummingbirds," he told me, "you see a design. . . . Why do we appreciate beauty? What accounts for the delicacy of flowers? Why is there such a remarkable interconnection?"

After accepting Christ, Randy initially never thought there was any conflict between his Christian faith and his love of the creation. But the more he talked with other Christians, the more he encountered attitudes and assumptions that gave him concern. In our conversation that day, as well as in others that followed, Randy identified four areas of questioning which he found among fellow Christians.

First, he sensed a "dualism" among many Christians. For them, life seemed to be divided between the body and the soul, between the material and the spiritual. When they die, he told me, these people know that their souls will go to heaven. They just happen to have this body now in this earthly life. But it is of no lasting importance to them. In the same way, the material things of earth don't matter compared to the spiritual things of heaven.

Second, Randy found that the term *world,* as used by Christians and as it appears in the Bible, seemed to be misunderstood. To most Christians, "the world" was equated with sin, and with what should be avoided. In their determination not to be "worldly" many of them assumed that any concern about the earth itself was simply to be avoided. But the biblical meaning of the term, Randy discovered, was far richer and more complex; God loved the world, and Christians are called to be free from sin, but not to withdraw from involvement in the world.

Third, Randy often heard Christians ask him, "The world is going to be destroyed anyway, so why worry about it?" Such Christians had strong views concerning the present plight of the world, but they had little hope for improvement until the Second Coming of Christ. Often, their attitude toward Randy's concerns for the protection of the earth proved so unsympathetic that any discussion was strained. For his part, Randy couldn't understand how belief in Christ's Second Coming removed the calling to live and act responsibly now, following biblical injunctions to address the problems found in the present world.

Finally, Randy struggled with others who felt, "Why do all these environmental problems matter anyway? Isn't saving people from sin the only cause we should be involved in?" But Randy believed that the Bible itself calls Christians to more than just preaching the gospel and that it presents a whole lifestyle for Christians to embrace. Any dichotomy between the social gospel and evangelism, in his view, wasn't possible. In fact, it seemed to Randy that involvement in issues within the world, such as protecting the environment, could be used in an evangelistic way.

At the university, Randy was actively involved in the work of Campus Crusade for Christ, a well-known evangelistic organization. He observed that many of the topics which his campus group discussed focused around personal issues, dating, and other concerns dealing with one's individual "walk with the Lord." Why, he asked, shouldn't this Campus Crusade chapter also express a biblical interest in other topics, such as the care of the environment? This seemed all the more appropriate since the university they attended featured a strong interest in forestry, wilderness, and environmental studies.

Eventually, Randy suggested to his local Campus Crusade director that they schedule a speech on the Christian's view of the environmental crisis and announce it campus-wide. The director agreed to this unique idea, and Randy spent much of the next six months in preparation.

Over the summer, he devoted many hours in biblical research. Beginning with the simple question, "What should a Christian's attitude toward nature be?" he used a concordance and tried to examine each relevant passage. His intention was to keep out his own personal bias and approach the Bible in an honest way, seeking to discover its own answers to that question.

Then he began reading books on the subject written by biblical scholars, theologians, and others. By the fall, he was organizing his research and answering the question of the Christian's attitude toward nature in a thoroughgoing way. He concluded that Christians must regard nature in light of the doctrines of creation, sin, and redemption and then determine their response. Since the basic call of the Christian is the submission of one's life to Jesus Christ, Randy concluded that this act should lead to following God's intentions for the protection of our God-given environment.

Randy gave his speech on a cold winter evening in February. The Campus Crusade chapter had publicized it throughout the university. Sixty people attended, some of them Christians wanting to know what the Bible had to say about the environment, some non-Christians who had the same question. This may have been the first time any Campus Crusade chapter had sponsored such an event. It proved to be both a new opportunity to expose non-Christians to the gospel, and to challenge Christians with the meaning of their faith for the care of the earth.

Randy's ongoing experiences and academic studies continued to underscore the need for Christian faith to be concretely related to environmental ethics. University courses on the philosophy and practice of conservation raised questions concerning basic values which, it seemed to Randy, were hard to answer apart from the foundation of Christian faith.

"Why, for instance, should we protect the Texas horned toad? Why should we keep it from becoming extinct?" he asked. Various attempts were made by environmentalists to give some basic value to such species apart from that of being an economic resource. But these attempts seemed to fall short, even by their own standards. Yet Randy was discovering that Christianity did provide a solid basis for giving value and protection to each species of life.

Randy also reflected on his experience in working with various environmental organizations. At one time he worked as an instructor for the National Wildlife Federation in special courses which they offered. His fellow staff members, who were mostly non-Christian, reflected a frustration in simply sharing information about the seriousness of various environmental problems. His staff colleagues and Randy knew that merely adding to the overload of facts in people's minds didn't result in different behavior. They realized the need to deal with matters at a spiritual level, rather than just intellectually. However, there was neither consensus about what that meant nor could a secular environmental organization such as theirs raise these deeper questions.

Randy graduated with a B.A. in zoology and a B.S. in recreation management. But how he would combine his Christian commitment with his environmental concerns remained unclear. He was caught in the dilemma of the insensitivity of the evangelical community to environmental issues, and the disinterest of the environmental community in Christianity.

As a starter, he took a summer job with a Christian ministry which worked with high school students and adults in wilderness experiences. Ultimately, he hoped to find some opportunity to work in outdoor recreation and environmental education in a Christian setting. But he knew that finding such a combination would be difficult.

When Randy and I last talked, he reflected on his experiences as a Christian and an environmentalist. When he saw others converted to Christianity at the university, if they had felt a strong environmental commitment before conversion, it often was maintained. But among those already a part of the evangelical community, he rarely found people who would make a decisive commitment to the environment as a result of their Christian experience.

Randy's story is not unique. During the last few years I met hundreds of people who had encountered similar experiences and frustrations. Those evangelicals with environmental concerns almost always feel isolated from their brothers and sisters in faith. Many who were raised with strong Christian commitments, and who then became involved in the struggle for environmental preservation, have rejected Christianity altogether because of the uncaring attitude of the church. And since the unleashing of the environmental movement in the past two decades, its adherents have usually regarded Christianity with caution at best, and often with private disdain.

Clearly, the Christian community has forgotten its God-given task to "tend the garden." Commitment to preserving God's creation has been nearly nonexistent, and often viewed as heretical or irrelevant by many sectors of the church. This has been true particularly for the evangelical community.

But another pathway is open for Christians to follow, a pathway clearly set forth in the witness of Scripture. Biblical faith gives humanity the responsibility of upholding God's purposes and intentions for the created order. Preservation and care of the environment should become for the Christian an extension of God's love, which creates, reconciles, and upholds all things.

Following this pathway means answering the questions which confronted Randy. In doing so, we will come to a deeper understanding of the full meaning of Christian faith. And we will also be able to express its relevance to a society which is both spiritually disoriented and is hastening the physical deterioration of life itself in God's creation.

2

All Creation Groans

In the late 1970s, an evangelical pastor in Puerto Rico founded Christian City. His desire was to build a "celestial community" that would be free from alcohol, gambling, drugs, and the other evils of urban life. Its low-cost housing, priced between $28,000 and $38,000, received financing through the Commonwealth of Puerto Rico's development agency.

The founder, Ray Figueroa, hoped to provide the residents of Christian City with a concrete expression of their new life in Christ. But shortly after moving into this Christian settlement, some residents became mysteriously ill. Terrible headaches, rashes, dizziness, loss of hair, frequent miscarriages, and other afflictions seemed to spread through Christian City like a plague.

Originally, the doctors could find no conclusive causes for these illnesses. Then, it was discovered that the fish swimming in Frontier Creek, which runs by Christian City, were contaminated with mercury. Tests by Puerto Rican environmental officials in 1985 showed surprisingly high levels of mercury on the banks of the creek and in the soil taken from those banks to fill the swampy area on which Christian City was built. Mercury taken in samples there proved to be from seventeen to thirty times higher than the average in the United States. Blood samples then taken revealed that two of every three residents of Christian City had higher than normal levels of mercury; many were suffering from its "intoxication."

During the 1970s, firms located upstream from the site of Christian City had regularly dumped industrial wastes in Frontier Creek. At least one of them, Technicon, which manufactured medical diagnostic equipment, produced mercury wastes. The direct cause-and-effect links between mercury pollution, soil contamination, and human illness are difficult to prove. But the realities are devastating.

In 1985, all twelve hundred residents of Christian City were evacuated. Further, because other citizens believed incorrectly that these people were carrying a contagious disease, the former Christian City residents—now living in temporary government housing—were often shunned and ostracized.

Elizabeth Algarin, a former resident, described her agonizing five years of life in Christian City to a reporter: "Not even in the Bible is there a hell as bad as this."

The Endangered Earth

This true story of Christian City is also a parable of the church's relationship to the environmental crisis. In their concentration on the spiritual realities of life in Christ, many Christians have been oblivious to the realities of this earth. But the "new creation" found in biblical promises can begin to emerge within this world only as the gifts of this creation are cherished and protected.

Failing to do so will lead to the creation of more hells on earth.

Never has the need for restoring the wholeness of creation been more urgent. The environment of air, water, earth, and fire (energy resources) sustains and nurtures life itself. As the environment deteriorates, life for us all is endangered. And when life-sustaining resources are degraded, eroded, or exhausted, life becomes threatened and lost.

Consider these realities:

- Each year on planet earth, an area of tropical forests the size of Scotland is destroyed and lost; soils then erode, the climate begins changing, and its replenishing resources are gone.
- Two-thirds of the forests standing in Nigeria in 1960 were eliminated by 1985. Liberia has lost 80 percent of its forests during the same period.
- India has lost an estimated 85 percent of its original forests; and if current deforestation rates continue, barely any forests will survive there by the end of this century.
- Thus far, this century has witnessed the elimination of approximately half of all the forests in developing countries.
- The lack of wood for fuel in poorer countries results in 400 million tons of animal dung being burned each year as household fuel—enough, if used as manure, to raise grain production by 20 million tons.
- Such tropical deforestation is the major cause of a modern mass extinction of plant and animal species as great as that which occurred after the disappearance of dinosaurs 65 million years ago.
- Scientists are estimating that as many as 1 million species of plant and animal life will become extinct, due to the human destruction of forests and ecosystems, by the end of the twentieth century.

- 850 million people living in the earth's dry land areas are threatened with desertification—the degrading of land caused by overgrazing, deforestation, poor irrigation, and rural neglect.
- Since 1977, five of the past years have been the warmest years our globe has experienced in this century.
- The increased burning of fossil fuels, the loss of forests, and other factors are warming the earth's atmosphere; by the middle of the next century the atmosphere could be six degrees (F) warmer, with potentially drastic effects on sea level, agriculture, and human health.
- Despite the banning of chlorofluorocarbons in aerosol cans in 1978, its other uses have increased, causing serious depletion of the atmosphere's ozone layer, threatening human health, and retarding crop productivity because of increased ultraviolet rays reaching the earth.
- According to the United Nations Environmental Program, twenty thousand cancer deaths in the United States each year may be resulting from radioactive radon that is accidentally trapped in well-insulated homes.
- The environmental quality in the developed world is seriously threatened by the chemical industry, as scores of human-made chemicals have entered into the food chain and are now found in the body fat of the population.
- Of the toxic chemical wastes produced in the U.S., only about 1 percent are actually destroyed.
- 77 percent of Americans, and 90 percent of their children under five, are carrying levels of lead in their bodies higher than that recommended by environmental health scientists.
- 70 million Americans are drinking water containing levels of lead higher than what is considered safe by the government's Environmental Protection Agency.

- Ten thousand people die each year from pesticide poisoning, and another forty thousand fall ill. The rapidly increasing use of such chemicals throughout the world threatens water quality and poses risks of increased cancer and birth defects.
- One third of all pesticides which are exported by the U.S. are legally banned for use at home.
- Total farm income in 1985 was $28 billion; the total expense of packaging for food that year was $29 billion.
- About one-third of all household garbage comes from food-packaging materials.
- In 1960, the average U.S. citizen produced 2.9 pounds of trash per day; today, 5 pounds per person per day go into the trash can.
- As garbage increases, landfills are being closed; 3,500 in the U.S. have closed since 1979. New York State had 1,600 landfills in 1965, and 518 in 1982. Some major cities now truck or ship nomadic garbage hundreds of miles in search of disposal.
- By 1990 the present landfills for half of the cities in the United States will be filled and unable to receive more garbage.
- As the Ogallala aquifer under the Great Plains of the U.S. continues to be depleted, the area of land irrigated in that region has decreased by 15 percent between 1978 and 1984.
- In India's southern state of Tamil Nadu, the water table has dropped twenty-five to thirty meters in the past decade.

A Godless View of Nature

These are but a few examples of global trends depicting the deterioration of the world's environment. While dramatic tragedies such as Bhopal and Chernobyl make headlines for a week or so, each day, in silent and often unnoticed ways, the gifts of life in creation are being poisoned and depleted. As the environment deteriorates, human wholeness is diminished.

At the heart of such problems lies a fundamental breakdown in the modern world view of Western culture. Since the Enlightenment and the scientific revolution, Western culture came to assume that humanity had both the right and duty to dominate nature. Objective, scientific knowledge became an absolute value. And the purpose of such knowledge was to exercise power over the creation. The view of life became secularized; we came to understand the world apart from any reference to God. The creation became "nature"—raw materials which existed only to be given value through exploitation.

The pragmatic benefits of these developments for civilization are remarkable. Several major diseases have been virtually eliminated from the earth. Communications and transportation have vastly enriched human experience. Humanity has been afforded protection against many ancient threats to life.

But today, this mindset is presenting humanity with more curses than blessings. Technology has become a social drug. We are addicted to technological solutions to any problem. Power seems the same as truth. Thus, we split the atom because we could do it. Instead of solving problems, that action gave humanity the godlike power of life and death over the created order.

Bhopal and Chernobyl are regarded as technical mistakes needing fixing, rather than as prophetic warnings to reconsider our mindless faith in humanity's mastery of nature. The Old Testament often describes God as "the maker of heaven and earth." But modern humanity, confident in its power to remake the earth, now begins, with "Star Wars," an effort to control the heavens as well.

At the heart of modern society, something has gone deeply wrong. We have become far too confident in our own power, and have trusted far too deeply in our dominance over the creation. We have constructed a world view which places human power and glory at the center of the universe. We have become like gods, masters over creation's destiny, and ready to demand any sacrifice for our enjoyment.

The Church's Challenge

And where in this picture is the voice and witness of the Christian church? Too often, it has served simply as a cheerleader for such "progress" rather than being the salt, light, and leaven essential for the world's preservation. Christians have blindly accepted the values and philosophy of modern Western society instead of searching for distinctive biblical perspectives regarding our relationship to God's creation and the search for true human wholeness.

Proof texts like "subdue the earth" have served as a sweeping rationalization for mindless exploitation of resources. The culture's determination to separate economics from ecology has gone largely unchallenged by the church. From the far right to the far left, economic views regard creation simply as raw material lacking any value until it is exploited and converted into a useful good. The intended biblical harmony between "tending the garden" or preserving the creation, and fulfilling human needs, is all but forgotten in modern practice.

The church's theology has failed to provide modern culture with an integrated vision of life. Evangelical theology has stressed personal conversion. More liberal theology has emphasized the movements in history to liberate people from oppression. Both have largely neglected the biblical emphasis on redeeming the creation. And neither has set forth an alternative vision for modern culture's understanding of humanity's relationship to creation, technology, and values in light of God's place as Creator of the heavens and the earth.

Continuing these trends risks a twofold disaster. First, the material basis for life itself, held in the gift of the environment, will continue to deteriorate and the actual integrity of the creation will be threatened. Second, the spiritual integration of life within modern culture will vanish, leaving society adrift—dominated by values that fragment people's lives and bereft of the spiritual resources essential for envisioning a hopeful future.

Fundamentally, modern Western culture needs an unshakable commitment to preserving the integrity of creation. Research into global environmental problems has mushroomed within the past decade. Our problem is not a lack of knowledge. Similarly, environmental advocacy groups which focus on particular issues have expanded their constituencies in impressive ways. Yet our culture as a whole fails to understand the clear threat to the foundations of life. Its underlying attitude and practices toward the creation have remained largely unchanged.

Advocacy around specific issues, while essential, is not sufficient. Political and regulatory reforms, while needed, do not reverse basic cultural attitudes. Legal strategies to protect resources, while helpful, do not transform values. Beyond all these measures, a basic conversion in our culture's relationship to God's creation is essential. Such a transformation would place the integrity of the creation at the heart of society's political, social, and economic activities.

In a similar manner, society stands in crying need of a unifying vision for human life expressed in practical ways. Rediscovering the intended relationship of humanity to God's creation will also result in a wholistic understanding of life, integrating the spirit with the mind and the body. Rather than hierarchical structures which enshrine humanity's enslavement of nature, as well as male domination of women, our culture yearns for models of partnership, stressing participation within the creation, as well as cooperation between the sexes.

For this to happen, the church must be reawakened to the vision from its own biblical tradition and must be renewed in the commitment of what its faith can offer culture and the creation. A theology of creation, stressing the relationship intended by God between humanity and the gift of the earth's resources, must break forth with strength and power within the church today. These truths, which are so strongly present within the biblical message, must be rediscovered and interpreted with relevance to today's church members.

In these ways, the church must break free from its captivity to Western culture. Christian faith will be renewed as it discerns the difference between truths rooted in biblical views and perspectives that merely reflect our culture's values. Increased relationships and serious theological dialogue with Christians from non-Western cultures will help church members in the United States understand more carefully the complex link between faith and culture.

The wholistic understanding of human life rooted in biblical faith needs to be freshly grasped and expressed by those committed to Christ. The church should come to understand itself as a model of new life, demonstrating an inviolate commitment to preserving the gifts of God's creation, and exhibiting an understanding of human life and relationships built on the wholeness intended by God. Thus, the church should serve the world as an expression of the new creation.

Within the United States, the church functions as the largest voluntary group in the society, with the ability to influence values and social commitments. Our culture's understanding of its purposes and goals is open to being influenced to some measure by the Judeo-Christian tradition. Therefore, if Western society is to transform its relationship to the creation, the church must play a critical role. It should provide the stimulus to challenge the culture's prevailing attitudes of exploitation and offer the culture a persuasive vision for altering its stance toward the creation.

The transformation of basic cultural values is a complex and arduous process. Yet, history demonstrates that people of faith, expressing and living out a vision within the society, can play a catalytic role. The abolition of slavery, women's suffrage, the civil rights movement, opposition to the war in Vietnam—all these had roots in small, grass-roots groups motivated usually by a deep religious faith that voiced an alternative moral vision to the society. Slowly, and through many acts of conscience, attitudes changed, laws were passed, and new structures emerged.

The challenge of preserving the integrity of creation and protecting human wholeness must eventually involve the gathered commitment of the world church, and countless other groups and organizations. At this stage, efforts toward the goal appear as mustard seeds. Yet, the call of the church to be salt, light, and leaven today directs it to be a catalyst in the task no less than saving the creation itself.

3

Why We Are in This Mess

The Mississippi River wants to change its course again. As always, it is seeking the shortest, steepest path to the Gulf. About every thousand years the mighty river's route through a couple hundred miles in southern Louisiana has shifted dramatically. As the soil and sediment from the midsection of the North American continent are deposited along the river's final entrance to the sea, its main channel eventually becomes a slower, more difficult route. So the river chooses another pathway, often at the time of exceptionally high water. Then a previously small outlet becomes the main course of the Mississippi's final journey into the Gulf of Mexico.

At about the time of the prophet Isaiah's birth, the Mississippi's main channel shifted from what is now the Bayou Teche to the east. Before the Nicene Creed was written, the river once again chose another course running to the south, which today is the Bayou Lafourche.

And a thousand years after Christ's birth, the Mississippi changed its course once more, running further to the east.

By the middle of the twentieth century, the river was ready to alter its course once again. Its easterly route had become lengthened, with the build-up of a delta. And a shorter, steeper outlet to the ocean was ready and waiting to capture the drainage of mid-America—the Atchafalaya River. Approximately two hundred miles northwest of New Orleans, where the Mississippi begins to form the border between Louisiana and the State of Mississippi, it connects with the Atchafalaya River. Following that river's course to the ocean is now less than half the distance of the Mississippi's present route. The mighty river, which just keeps rollin' along, wants to go that way.

But in this last millennium, white settlers arrived in North America. New Orleans and Baton Rouge have grown up alongside the river. Industries between those cities— many of them large chemical and petroleum plants—have been drawn to the fresh water and shipping afforded by the Mississippi. Those cities and industries could not tolerate nor endure a natural decision by the river to change its direction.

So, war was declared on the Mississippi, with the Army Corps of Engineers leading the battle. A film the corps made describing some of its early efforts at preventing the capture of the Mississippi by the Atchafalaya included these words:

> This nation has a large and powerful adversary. Our opponent could cause the United States to lose nearly all her seaborne commerce, to lose her standing as first among trading nations. . . . We are fighting Mother Nature. . . . It's a battle we have to fight day by day, year by year; the health of our economy depends on victory.[1]

The connection between the Mississippi and the Atchafalaya is called Old River. There, the Army Corps of Engineers carries out its main battle plan. The centerpiece of its strategy is a creation called the Old River Control

Structure, which attempts to allow 30 percent of the Mississippi's flow to go through to the Atchafalaya, but no more. This was the approximate distribution in 1950, and Congress passed a law decreeing that it should forever remain the same.

Major General Thomas Sands, a veteran of Vietnam, is directing this battle. "In terms of hydrology, what we've done here at Old River is stop time. We have, in effect, stopped time in terms of the distribution of flows. Man is directing the maturing process of the Atchafalaya and the lower Mississippi," he explains.[2]

But how long the U.S. Army can stop time seems unclear. In the flood of 1973, when high water was only 20 percent above normal, the Old River Control Structure was shaken to its core. The Army Corps' own report from this battle revealed that 50 percent of the foundation under the structure was eroded. Emergency reinforcements were called in to narrowly prevent the victory of the Mississippi and the Atchafalaya over the Army.

Yet, in subsequent testimony to the Congress, a general assured concerned legislators that "the Corps of Engineers can make the Mississippi River go anywhere the Corps directs it to go."[3] In southern Louisiana, the Corps, in the words of one author, "has been conceded the almighty role of God."[4]

Many observers of this battle are convinced that the rivers inevitably will win. Raphael Kazmann, a retired hydrologic engineer at Louisiana State University, put it this way:

> Floods are more frequent. There will be a bigger and bigger differential head as time goes on. It almost went out in '73. Sooner or later, it will be undermined or by-passed—give way. I have a lot of respect for Mother . . . for this alluvial river of ours. I don't want to be around here when it happens.[5]

And the *Washington Post*, commenting on this conflict, editorialized in 1980: "Who will win as this slow-motion confrontation between humankind and nature goes on? No one really knows. But . . . if we had to bet, we would bet on the river."[6]

Yet, General Sands remains determined to hold the fort at Old River against the onslaughts of the Mississippi. "Man against nature. That's what life's all about," he declares.[7]

At War Against Creation

This militaristic attitude of being at war with nature is embedded deep within Western culture. The battle being led by General Sands and the U.S. Army against the mighty Mississippi provides us with the paradigm that has governed the actions of the modern West, and, in fact, most of the modern world. And in this, the Christian church has served as a loyal chaplain to the troops. We have blessed the battle with suitable proof texts, such as "subdue the earth and have dominion." And we will pray for victory by asking God not to send a heavy rain to the Mississippi Valley, even though God has seen fit to do so regularly for the past several millennia. When the rains and the flood do come, the church will be there with the Red Cross, just as in other wars, giving relief to the wounded. But the church will not have raised its voice and witness against the war itself.

How have we gotten into this mess?

Modern culture has been waging a war against nature, and the ecological foundations for sustaining the earth's life are eroding. Yet, the Christian church has remained compliant and silent. If we can understand some of the causes behind this cultural and theological tragedy, perhaps we can find the directives that will lead the church, and point the way for the culture, in establishing peace—God's peace—with the creation. Then it may become possible to learn how to live in peace with rivers that are flowing into oceans.

At the outset, it is important to discard misguided explanations for our plight. Chief here is the theory that both the environment and the church are in this ecological and theological predicament because of what the Bible says. Lynn White, Jr., gave us the most eloquent and popular expression of this view two decades ago in his well-known article, "The Historical Roots of Our Ecologic Crisis" (see the Appendix). There, he argued that the biblical injunction to "subdue the earth" in Genesis 1:28, and the destruction of the idea (as a result of Christian teaching) of any sacred presence within nature, opened the way for the modern scientific and technological assault on the natural order.

White's article blaming Christianity as the chief cause of the ecological crisis quickly gained the status of inerrancy among many environmentalists. But the past twenty years of discussion have clarified matters. First, White's description of biblical teaching regarding the environment is selective and highly distorted. Second, his argument that Christianity paved the way for the scientific and technological revolutions is very questionable. And third, his assumption that environmental destruction has flowed solely from the mindset of Western culture, and not from others, is historically dubious.

We can be grateful to Lynn White for starting afresh the debate about Christianity and the environmental crisis. But it is high time that we move beyond the assumption that the ecological devastation of the creation has been caused by a few words of misunderstood Scripture. As René Dubos has written, "If men are more destructive now than they were in the past, it is because there are more of them and because they have at their command more powerful means of destruction, not because they have been influenced by the Bible."[8]

The Roots of Our Problem

Can we, however, identify some reasons that explain afresh our present ecological and theological plight? I think so. Let me suggest several.

First, Christian faith in the West has been captive to the assumptions of modern culture which sever God from the creation and subject the creation to humanity's arrogant and unrestrained power. Unlike Lynn White, I believe our problem lies in the church's historical captivity to Western culture, rather than the reverse. One cannot underestimate the way in which the Enlightenment and the scientific revolution, and the process of industrialization which followed, dramatically altered humanity's relationship to the physical environment.

Bacon, Descartes, Newton, and others set forth a view of nature as raw material, governed by laws contained wholly within itself and existing for the sole purpose of being exploited. Political and economic theorists like John Locke and Adam Smith further explained that nature only had value when it was turned into something useful. The plunder of the earth became justified by the pursuit of individual freedom, knowledge, and prosperity.

But something more happened. The mechanical, mathematical materialism that developed from this revolution constricted the arena in which truth could be known, and for certainty to be established. Now, reality could be proved rather than accepted by faith. The true nature of the world could be known through the scientific method, and such knowledge would enable human progress.

For this to happen, God's relationship to the creation had to be severed in the understanding of the modern mind. The idea of God had been used by the church, often in a foolish, pig-headed manner, to explain the world. But now better, more convincing, more verifiable explanations were becoming possible. So "nature" had to be understood as an object unto itself, apart from its relationship to God. In other words, it became secularized.

For the purposes of the scientific method, and for the applications of technology which followed, God was simply irrelevant. Moreover, the materialist view of reality continually sought to break the world down into its smallest component parts. Any so-called "spiritual" dimension of reality was, by definition, outside the realm of observable fact and thus not "true," according to this view of truth.

Today, the views of truth and reality which began the scientific revolution are under critical reexamination, most notably by scientists themselves. Others are discarding this worldview wholeheartedly. The relationship of the parts to the whole is more important, for instance, than the understanding of cause and effect. And the spiritual dimensions of the creation, which link it together as a whole, are being proclaimed as evident realities, essential to grasping the reality of the world.

This makes our present age at once exciting, challenging, and confusing. Beyond doubt, we are in what can be called a "paradigm shift," where the culture is questioning its former certainties about the world, and seeking new models for understanding the cosmos. Ironically, many evangelical Christians find this all very threatening, and have become paranoid about what they call "New Age" ideas. Apparently, such Christians feel more secure with views of reality which were actually constructed to make belief in God irrelevant to the discovery of truth in the world.

Second, modern cultural and theological assumptions have placed humanity at the center of purpose and meaning in the universe. Questioning this perspective seems almost unthinkable to us. Yet, a decisive evolution in modern thought has concluded that humanity's autonomous goals and purposes are the chief end of the creation. The entire modern concept of individual "rights," which forms the basis for Western political and economic ideologies, is grounded in this belief.

Within this framework, any question of values or ethics is decided on the basis of the effects on humanity's welfare. For instance, the value of any given species in the creation is dependent upon its relationship to humanity. And the environment as a whole exists to serve and sustain human life.

Theologically, the church is caught in this same trap. Particularly since the Reformation, and from the dominant influences since the Enlightenment, most modern Western theology assumes that the relationship of God to human experience, either personally or historically, is the center of the theological task. Our theology as well as our cultural assumptions revolve around an anthropocentric perspective.

But the truer task of theology, it seems to me, is to understand how the world is related to God. One theologian, Julian Hartt, describes theology's job in this way: "It means an intention to relate to all things in ways appropriate to their belonging to God."[9] This, of course, includes human experience, but not in a way that regards humanity in isolation from its inherent relationship to the rest of the creation. And creation itself is understood through its relationship to the Creator.

The common modern assumption is that God created the world for humanity's benefit. But the view echoed more persistently in biblical passages is that the creation exists for God's glory. In his provocative book, *Ethics from a Theocentric Perspective,* James M. Gustafson says this:

> . . . the purposes of the divine governance are, insofar as human beings can discern them, not exhausted by its benefits for us. Whatever they may be directed toward, they are not all directed toward us. This perception is the appropriate basis for acknowledging the transcendence of God, the ineffable mystery before which we stand in silence.[10]

Right-wing fundamentalists today are constantly warning about the dangers of "secular humanism" infiltrating our schools and our media. Other Christians are so appalled by their tactics and proposals that their complaints are never seriously evaluated. But in fact, modern Western culture has enshrined secularism—the understanding of life apart from any reference to God. And our culture is humanistic—believing that human goals and purposes are the center of meaning in the universe. While

I forcefully reject the solutions which fundamentalists propose, their rhetoric points to a deep truth concerning our cultural assumptions.

Reversing the anthropocentrism of modern culture and theology requires a wholesale reconstruction of how we tend to think. Yet, to be freed from our present plight will require such a reorientation in how we approach the creation, in order to better grasp God's glory. But is this not what Paul urged when he wrote in Romans:

> Adapt yourselves no longer to the pattern of this present world, but let your minds be remade and your whole nature thus transformed. Then you will be able to discern the will of God, and to know what is good, acceptable, and perfect. (Rom. 12:2 NEB)

Third, our culture adheres to a blind faith in technological progress as the means to resolve environmental problems and the maldistribution of world resources. Bhopal and Chernobyl should serve as this decade's prophetic warning signs. The technological idolatry of the modern world must be resisted in the same way that Old Testament prophets opposed the haughty arrogance of Israel's unfaithful kings.

Early on a December morning in 1984, about a quarter of a million people in Bhopal, India, mostly poor, found themselves in a gas chamber. At least two thousand died, and tens of thousands more suffered ongoing serious injuries. *The Times* of India wrote, "If there was a wretchedly undignified, hideously helpless form of megadeath after Hiroshima and Nagasaki, this is it."[11]

At Bhopal, the Union Carbide Company manufactured pesticides, using a highly toxic chemical, methyl isocyanate. In the aftermath of this tragedy, the investigation centered around what had technologically gone wrong, and how that could be remedied. But another, more radical view would identify the problem as simply the manufacturing of such pesticides.

A history of leaks and accidents has characterized the

operation of Union Carbide plants both at Bhopal and at Institute, West Virginia, even before and after the 1984 tragedy. This is not because Union Carbide is irresponsible in its safety procedures, or has failed to search hard for means to guard against such accidents. It has been as vigilant as most companies engaged in this business. Rather, the process of manufacturing these pesticides is inherently risky and dangerous, regardless of how careful one tries to be.

Instead of simply trying to make such manufacturing more foolproof technologically, the use of such pesticides should be made less necessary. A framework of values protecting the integrity of the creation, and political structures able to implement them, must guide the kinds of technological choices made in modern societies. Instead, we have largely allowed technology as an end in itself to determine such choices. In the process, creation is exposed to potentially devastating dangers.

Likewise, the incident at Chernobyl should not be regarded as an "accident." Rather, this is an outcome of a widespread reliance on nuclear energy. As energy analyst Amory Lovins explained, "The deeper lessons are not technical, they are human. . . . The reason I call it an inevitability instead of an accident is that if you put megaton inventories of fallout inside bottles, sooner or later somebody will screw up enough to let it out."[12]

Beyond such dramatic outcomes of nuclear technology, like Chernobyl, is the threatening presence of fourteen thousand tons of high-level waste which must be absolutely contained for ten thousand years. As Lovins says, such time periods are "more theological than geological." The nuclear industry has proceeded in the faith that this technological problem will yield to a technological solution. But it has not.

A wise approach to these realities would be for a nation to decide that its reliance on this technology is a case of mistaken faith. Sweden has done so, committing itself to end its reliance on nuclear energy, making a transition to other alternatives in a reasonable time. Such a decision, however, necessitates using a basis of values to guide the technological choices made by a society.

Such actions are less heretical than we might initially suppose. In fact, those areas where positive gains in environmental quality in the United States have been made in the past two decades involve decisions to ban certain substances and related technologies. As Barry Commoner has wisely pointed out, those pollutants which have been reduced most dramatically are "lead, DDT and similar chlorinated pesticides, mercury in surface waters, radioactive fallout from nuclear-bomb tests, and, in some rivers, phosphates."[13] In all these cases, the technological process producing the particular pollutant was altered or stopped. Lead is being phased out of gasoline. Phosphates have been banned. Atmospheric testing of nuclear weapons has halted. Mercury is no longer used in the production of chlorine along the shores of Lake Erie.

On the other hand, when faith has been placed in technological controls to limit the amount of pollutants entering the environment, the results are far less encouraging. While such controls certainly make a difference—scrubbers in smoke stacks and catalytic converters on cars, for example—such differences have not substantially altered, much less solved, the presence of such pollutants within the environment. Acid rain continues, and cities are seriously affected by poisonous smog.

Commoner's conclusion seems instructive:

> . . . pollution levels can be reduced enough to at least approach the goal of elimination only if the production or the use of the offending substances is halted. This precept directs our attention to the technology of production. . . . all the really successful environmental improvements have been achieved by altering that technology.[14]

To repeat, choices guiding technology must be implemented through a framework of values committed to upholding the integrity of the creation. Here, the church should play a critical role in helping to formulate such values and advocate their practical application.

In no area will this need be greater than in biotechnology and genetic engineering. Here, society is witnessing fundamental choices being made concerning the integrity of the creation and its relationship to humanity. The recent ruling of the U.S. Patent Office, allowing the patenting of genetically altered animals, is a startling example. This ruling, if implemented, allows all of animal life to be regarded as merely a commercial commodity, subject to unrestrained manipulation. New species, in effect, can be created and then marketed, with royalties paid to companies on each of their offspring.

A fundamental respect for the integrity of species is erased. Humanity's right to genetically change the actual forms of created life is condoned, simply because it is technologically possible and economically profitable. Here we see illustrated most vividly modern society's addictive faith in technology, and the urgent need for values rooted in spiritual realities to direct the functioning of technology within the created order.

Fourth, the Western church's modern theology has fought between being personalized or politicized, and largely has forgotten the theology of creation as its starting point. In the theology of creation, the place of the whole created order is given prominence. Thus, God's presence as sustainer of the creation comes before God's mighty acts in history. Covenant begins with the story of Noah, and is made between God and all the earth, rather than beginning with the promises to Abraham. The mark of faithfulness or disobedience toward God for the people of Israel finds expression in their relationship to the land—is it God's gift, to be shared, and or a possession, to be held?

The incarnation of Jesus Christ is seen by the theology of creation as God's vulnerable embrace of the whole creation. Salvation and redemption, then, are related not just to individuals, but rather to the creation itself. God's intended purposes for the world are promised fulfillment;

Jesus Christ is the life of the world. Therefore, the new creation, though always an eschatological hope, finds its initial expression within this creation, and not beyond it. The Spirit, who was present at the first act of creation, and creates the church, is poured into the world, "renewing the face of the earth." Views of sin, salvation, mission, service, and future promise all are formulated with a focus, biblically derived, on the relationship of creation to God and to humanity.

But these theological perspectives have been largely neglected in the fight between those who would merely personalize the biblical message and others who would merely politicize it. Reducing the truth of Scripture to solely an individualized faith has long been the temptation faced by evangelicals. In many respects, evangelicalism as we know it today could not have flourished prior to the gradual focus of Western culture on the primacy of individual human experience.

The experience of conversion—a person's encounter with the Good News of God's grace through Jesus Christ— must always remain central to Christian faith. But when faith is confined to measuring one's inner spiritual temperature, with a wall erected between the spiritual and the material, then the full power and scope of God's grace in Christ is simply denied. In such a mindset, environmental problems at best are nothing more than another issue over which Christians may have different opinions, all largely unrelated to the gospel.

The politicalization of Christian faith has come largely in response to the failures of an individualized gospel. Quite understandably, the social, political, and economic dimensions of the biblical message have been stressed as central to the church's faith and witness. Moreover, the church has been thrust into the midst of the struggle of the poor for justice.

Liberation theology has become a major preoccupation of the Western church in recent years. Generally speaking, this approach insists that truth of the gospel is known only in the "praxis" or actual experience of the people's historical struggle against oppression. Such struggles are seen as holding the signs of God's promised coming kingdom. Often, Marxist ideological categories are employed both to analyze existing economic realities and to provide an alternative lens for interpreting the message of the Bible.

The gross injustices against which such theology is formulated should be outrageous to any follower of Jesus. And the past complicity of the church in cooperating with these oppressive structures is an offense to the gospel. Yet, liberation theology remains limited in many respects.

In its insistence that God's work is discovered only in the historical struggle, the foundation of creation is simply ignored.

In its focus on the people's struggle, faith can remain centered solely on human experience. The anthropocentrism of theology, it seems to me, is often underscored just as strongly by liberation theologians as by television evangelists. And even in its search for an alternative ideology, liberation theology remains captive to Western perspectives which are convinced of humanity's power over history and creation.

The challenge remains for the theology of creation to demonstrate how it offers the church an understanding of humanity's relationship to God which is fully personal, while not individualized. At the same time, such theology must demonstrate that its vision of the biblical message is, in fact, fully liberating and a basis for overcoming all oppression through the promise of the new creation. This becomes the case, I believe, once we recognize that Christian faith directs us to God who is first Creator and then Redeemer, and to Jesus who is Savior and Liberator, but also Lord over all the earth.

Fifth, the Western church has been theologically arrogant and inattentive as well as condescending toward non-Western Christian perspectives. The economic and political colonialism of the West was accompanied by the theological colonialism of the church. And we are still suffering from its consequences.

At a conference on Asian theology held in Kyoto, Japan, I listened to the president of a distinguished Christian university and seminary in Indonesia explain his personal experience as a candidate for ministry facing ordination examinations in the Indonesian Reformed Church. He was asked to sign a statement saying that the Heidelberg Catechism is the most faithful theological expression of Christian faith. And he refused. My Reformed Church friend in Indonesia was not against the Heidelberg Catechism. But he could not understand why a theological confession written in Germany in the sixteenth century should be the final word on theology for a Christian living in Indonesia in the twentieth century.

A well-known Christian theologian from Asia, Dr. C. S. Song, has stated that young theologians in Asia today need a "vaccination" against Western theology. The truth of the gospel for Christians in Asia today will be discovered fully only as the cloak of Western culture is removed, allowing the biblical message to come alive with fresh vitality within those cultures.

One of the distinctive features of Asian theology, as it emerges, is the place of creation in their understanding of faith. Undoubtedly, this comes in part because such Christians are living within cultures that historically have not drawn such a sharp separation between humanity and the environment. These Christians are no longer convinced that a so-called "Eastern view" of the close interrelationship between God, humanity, and the creation should be discarded in favor of a "Western view" because it claims to be more biblical. They are perceiving, correctly, that the actual question is not what the Bible says, but rather, how Western culture has interpreted biblical teaching.

One of the most helpful ways for Christians in the West to gain fresh biblical perceptions of our relationship to the creation is through dialogue with Christians from non-Western cultural traditions. After centuries of attempts by the Western church to interpret Christianity to Eastern, Asian cultures, for instance, today it is time for Christians from those cultures to interpret Christianity to us. To do so honestly, however, will require at least a suspension of the West's theological bias that its understanding of Christian faith is the truest.

From its very beginning, as Christian faith spread westward, its converts, in turn, saw heresy to the east. From Jerusalem, the church's dominant outreach looked west. When power shifted to Constantinople, everything to the east was heretical. When the church split and power was consolidated by the bishop of Rome, everything east of there, including Eastern orthodoxy, was heretical. With the Reformation, everything east of Geneva was heresy. And today, it seems, everything east of Chicago has fallen into heresy.

Perhaps the westward flight of the Western church to escape Eastern heresy will eventually land us backwards, across the Pacific in the Far East. If we then could turn around and listen to our brothers and sisters in Christ, we might discover truths of the gospel which have been hidden by our arrogant and paranoid flight from the east. Like Jonah encountering Nineveh, perhaps we will sense that it is God's full truth we have wished to avoid. Maybe we, like him, will end up perplexed over why God has not judged in the way we think God ought to judge. And then we might sense anew the meaning of God's providence, and the reach of God's grace in Jesus Christ.

This experience may also help us rethink our attitudes toward the indigenous cultures within our own continent which were destroyed by the coming of Western culture. When the Spanish, English, and French explorers were bringing that culture to the shores of North America, some five hundred different cultural groups, with special understandings of humanity's relationship to the natural world, were thriving. Are we today to believe that God's grace was totally withheld from them, and present only with the destroyers of those cultures?

While St. Paul was traveling west, tradition holds that

the Apostle Thomas traveled east. Near the airport in Madras, India, is Mount St. Thomas, said to be the site where Thomas was martyred. Church history, at this point, fades into story and tradition. But it is a fact that the Christian church has existed in southern India since the third century. It was, of course, not the West's church; that came only with the missionaries. Yet, the Christian church was there.

Our Western biases have left us inattentive to the whole presence and witness of the Eastern church in the world. Eastern Orthodoxy, with its peculiar fascination with liturgy and seeming detachment from history, feels irrelevant to the West. And yet, its theological understanding of creation and redemption, which for centuries has celebrated God's action in Jesus Christ as reconciling the whole cosmos to God, has much to offer our theological search today.

Christianity, after all, began as a Near Eastern religion. In our desire to respond to both the ecological crisis and our theological failure, the church would do well to reconsider the Eastern roots and branches of its faith. When we realize that Western Christianity is simply one branch of this tree of life growing since the beginning of time in the world, we may better understand how God's promises are to find fulfillment.

The vision of Revelation tells us that the tree of life bore twelve kinds of fruit, and that its leaves were for the healing of the nations. The present ecological crisis has occurred in large part because modern culture believes it can thrive apart from the presence of this tree of life. And the theological failure of the Western church has come largely because we have assumed our theological fruit is the only one tasty enough to eat and to nourish us.

Let us look again at the tree on Calvary, on that hill in the Near East, and see once more the outpouring of God's love to all the earth. And then let us live in the promise of the tree of life, discovering its branches and fruit covering the earth, offering the healing and wholeness intended by our Creator.

4

Biblical Wisdom

In the summer of 1987, five hundred people gathered from all parts of the United States and beyond for the North American Conference on Christianity and Ecology. The gathering was the first of its kind, bringing together a highly diverse group of people. All shared a common concern for Christianity and the ecological crisis.

Dr. Calvin DeWitt, director of the AuSable Trails Environmental Institute, an evangelical teaching center for college students promoting Christian stewardship of the earth, was vice-chair of the board that organized the conference. He was an active participant along with other evangelicals. Father Albert Fritsch, a Jesuit who has long been active in justice and environmental efforts in Appalachia, served as the board's chair. Representatives from mainline Protestant denominations, the Orthodox Church, the Catholic Church, and peace churches all took part in organizing this event.

A vast array of speakers included Hans Schwartz, a German theologian; Wendell Berry, a poet, writer and farmer; Jeremy Rifkin, author and activist; and Thomas Berry, whose writings have challenged Christianity to express the "new story" of the universe.

Theologies of every sort were represented. Those influenced by the Institute for Creation-Centered Spirituality, for instance, argued that the traditional categories of Fall and Redemption have done much to spawn a degrading view of the creation. Following the view of that Institute's founder, Matthew Fox, they urged that all emphasis be placed on the goodness, blessing, and revelation of creation. Those elements in the Bible and those individuals in the Christian tradition who heralded that message should be affirmed, they urged, while that which portrayed a different view should be neglected.

Some went to even further extremes, echoing the view of Thomas Berry. Christianity, they argued, along with other great religions in the world, has played a role up until now in interpreting the creation and human experience. But today we must turn first to the story and primary revelation of the universe itself, and learn from it. In so doing, historic Christianity will no longer be adequate. It will need to be completely transformed in order to join in expressing this "new story."

"It would do no good," asserted Berry, "to impose extrinsic solutions derived from scriptural traditions of the past. The story of the universe as we know it is itself revelatory of that numinous mystery whence all things emerged in the beginning. . . . "[1]

The widely divergent views expressed at the gathering became polarized when the conference was asked to approve a statement for circulation among churches. For several months, a drafting committee appointed by the board of the conference had worked diligently on this proposed document. It began with a prayer of thanksgiving to God for the gift of creation, and confession over humanity's abuse of this gift. Then it set forth a theology of creation and redemption, summarizing central teachings of Scripture with reference to various texts. It concluded with a section on the ethics of faith and action.

As the conference met in plenary session to consider the draft document, the polarization became evident. A minority, including Thomas Berry, registered strong objections to the proposed statement because it used biblical references as a foundation. They objected to setting forth the Bible as the starting point, and argued that this established a limited and ineffective means for responding to the ecological crisis.

Many of the participants were genuinely perplexed by objections to the use of a biblical framework for addressing the ecological crisis. The conference, after all, was on Christianity and Ecology, and the intention was to reach the local church with the task of environmental preservation. Theologically, the appeal to see the creation, in and of itself, as the primary revelation of God struck many as a new version of an ancient theological problem— removing the distinction between the Creator and the creation. Pragmatically, many doubted that appeals for listening to the universe tell the new story of its numinous mystery would be very persuasive to an adult Sunday School class in most any church.

The minority conviction that biblical perspectives rooted in the Christian tradition were inadequate for addressing the gravity of the ecological crisis prevented the conference from reaching agreement on the proposed statement. Eventually, the board decided to make the statement available to interested groups and churches. (See page 182 in the Appendix.) It remains a valuable resource for Christians seeking to relate their faith to the crisis of the earth's environment.

In light of the conflict, when my turn came to give my address I chose to present the case for the strong mandate present in the Bible for the preservation of the earth's environment. After several biblical illustrations and references, I asked for questions.

The first came from Paul Ryan. Earlier in his life, Ryan had been in training with the Passionists, an order in the Roman Catholic Church. Leaving that order around the time of Vatican II, he had spent the next twenty years working with ecological issues and communication, particularly innovative ideas for the use of television.

Ryan strongly challenged my presentation. In all his years of working with ecological issues, the church had never been of any help. Rather, it was the church's and the Bible's ideas which he had combated. He argued, didn't the Bible, especially in the Old Testament, picture God as remote, powerful, and vengeful, ready to send forth floods and destroy mountains in order to carry out his wrath? How can we expect the Bible, with stories such as these, to be a source of inspiration for ecological responsibility, he asked.

In the dialogue that followed, I maintained that the church's past neglect of the environment, as well as the flagrant misinterpretation of some biblical texts, should not cause us to reject the biblical tradition as incapable of addressing the crisis of the creation. And if our goal is to reach the church with the message of Christian responsibility for the care of the earth, then it is essential, I said, to understand the relevance of the Scriptures for this task. Paul Ryan concluded that he hoped I could succeed in this effort, but he honestly doubted that it was possible. He believed that the Bible was more the problem than the solution.

A few weeks later, a woman from the local Missoula, Montana area called me at the New Creation Institute. Having previously expressed curiosity about our work, she had received literature describing the concerns, programs, and goals of the Institute. She began by asking about our name, and the biblical reference, 2 Corinthians 5:17, from which it comes.

"I don't see anything in this verse or in my Bible about the creation being made new. This applies only to the individual. You're mistranslating the Scripture," she began.

I tried to explain the meaning of the original Greek in that passage and in others. But it seemed as though she was calling not to ask questions, but to deliver her conclusions.

"God made man in his image," she went on. "The flowers, the mountains, and the land—these weren't made in God's image, only man. And man sinned. So man is in need of redemption, not the creation."

Once again, I sought for some opening for dialogue, asking her if she really was interested in what the Bible had to say regarding our responsibility to care for the creation. The effects of sin, I said, result in the destruction of the environment, and the work of redemption is meant to include the restoring of our relationship to the creation as humble caretakers of God's gifts.

"No, you've got it wrong, " she announced. "The Bible is concerned only about the individual. Salvation of the person through Christ is what the Bible's message is all about. There isn't anything I find in my Bible about the environment. I've been a Christian for over fifty years, and I know what the Bible says."

I was late for dinner. My patience wearing thin, I offered to send her material which she could study concerning the relevance of passages from the Old and New Testaments to the call to care for the earth. But instead, she moved from disagreement to judgment.

"I recently heard a program on Christian radio warning about the dangers of the New Age movement," she told me. "All kinds of non-Christian ideas are filtering into the church these days, and we have to be on guard and very careful. Talking about the environment and the earth sounds like one of those ideas . . . ."

Why is it, I thought, *that my anger can be so aroused toward fellow Christians?* Trying to control my feelings, I said plainly that we had nothing to do with the so-called New Age movement. We were rooted in Christ, and are called to alert the church to the full meaning of Christian faith, especially in light of the global ecological crisis. But by now, the conversation had ended. She had delivered her conclusions and her judgments, and I arrived home for dinner thoroughly frustrated.

Later, I reflected on these two encounters. Paul Ryan and this fundamentalist woman (she never did give me her name) shared a basic conviction that the Bible had nothing relevant to say about the environment. Though their reasons and experiences were completely opposite, their conclusion was the same. For one, the ecological movement had destroyed faith in the Bible. For the other, fundamentalist faith had destroyed all trust in the ecological movement.

These two encounters merely symbolize the widespread skepticism from the left to the right concerning the relationship between the biblical tradition and environmental protection. In countless other conversations with environmental activists, seminary students, mainline Protestant ministers, and committed evangelicals, I have found confusion, neglect, and simple biblical ignorance when it comes to the Scriptures' message concerning the creation.

The need for fresh attention to the whole of the Bible's perspectives on the earth, the land, the soil, animals, water, food, the environment—in short, the entire created order—is urgent. But thankfully, the last five years have witnessed a renewal in such biblical study.[2] Topics long neglected, or simply not noticed, have begun to receive careful attention. And the list of solid writings on the theology of creation and care for the earth has steadily grown.

As always, it takes time for the recovering of such biblical perspectives to filter down to the pulpits, much less to the pews of our churches. But this process has begun. So let us consider a survey of those passages and perspectives from the Bible which are beginning to receive the attention of a growing number of Christians today.

The Goodness of Creation

The whole creation, according to the Bible, renders praise and gratitude to God. Unlike anthropocentric perspectives which assume that the only purpose or value for the environment is to satisfy human need, the biblical view assigns goodness to the creation simply because it is the work of God. Further, all parts of the creation, not just humanity, are pictured as praising God's glory with thanksgiving and joy.

The goodness of the creation is underscored in the first chapter of Genesis. "And God saw that it was good" (which can also mean "right," or "well-ordered") are the words continually repeated after the appearance of water, land, plants, light, fish, birds, and land animals. And the finished creation as a whole, including humanity, is seen as "very good."

The creation, however, is never pictured as a detached entity, set apart from God and left on its own. Rather, the whole creation responds to the Creator. The Psalms constantly echo this theme. In Psalm 96, for example, all the earth "sings to the Lord" (v. 1). And the psalmist then declares, "Let the heavens be glad, and let the earth rejoice; let the sea roar, and all that fills it; let the field exult, and everything in it! Then shall all the trees of the wood sing for joy before the Lord . . ." (vv. 11–12).

In Psalm 148, every aspect of creation is enjoined to give praise to God. The sun, moon and stars, fire and hail, snow and frost, mountains, fruit trees, cedars, beasts, cattle—all these praise the Lord from the heavens and from the earth. Repeatedly, we read that the earth itself rejoices because of the Lord's reign. Psalm 97:1 goes on to add, "Let the many coastlands be glad!" In verse 7, the heavens proclaim God's righteousness, a sentiment echoed in Psalm 19:1, "The heavens are telling the glory of God."

Scriptures portray God as intimately involved in the ongoing life of the creation. One dramatic example is Psalm 104, which describes the dependence of each part of the creation on the upholding and sustaining presence of God. Toward the end of this marvelous picture, the psalmist's prayer is, "May the glory of the Lord endure for ever, may the Lord rejoice in his works" (v. 31). Thus, the creation even becomes the place of God's own rejoicing.

The Scriptures are full of God's actions as the Creator and sustainer of the heavens and the earth. Only plain biblical illiteracy can explain why many Christians assume that all the important passages on creation are limited to the first three chapters of Genesis. Many passages portray the "word of the Lord" as bringing forth the creation, such as Psalm 33:6–9. This becomes a foundation for the declarations of John, describing Christ as the Word through whom all things were made (John 1:3). Likewise, passages also picture God's breath, or Spirit, as upholding all created life (Psalm 104:30). This image becomes even richer when we realize that the same Hebrew word (*ruach*) means both spirit and breath, as well as wind. Thus, in Genesis 1:2, the Spirit, or wind, or breath of God moves through the dark, formless void over the face of the waters and brings forth the created world.

Words of Wisdom

One whole portion of the Bible particularly rich in the theology of creation is its "wisdom literature." Biblical scholars generally identify this as a whole tradition, or voice, present within the Old Testament, and flowing into the New Testament as well. Certain books of the Old Testament fully embody this perspective, including Proverbs, Job, Ecclesiastes, and various Psalms. Students of the Old Testament frequently see other portions as well reflecting this wisdom tradition.

In general, wisdom literature focuses on the creation, including human experience, in an open search for God's truth. And it then seeks to order life according to the truth that is discovered. In this way, the guiding, nurturing presence of God is revealed. That is why much of wisdom literature consists of observations and teachings drawn both from human experience and knowledge of the creation. Proverbs 3:19–20 states:

The Lord by wisdom founded the earth;
   by understanding he established the heavens;
by his knowledge the deeps broke forth,
   and the clouds drop down the dew.

Such passages see God's wisdom both as the source of creation and as reflected throughout the environment. In the twelfth chapter of Job, we hear Job argue that any of the beasts, birds, plants, or fish can teach his friends wisdom, for all these know what the hand of God has done. "In [God's] hand is the life of every living thing . . . " (12:10). And when the famed wisdom of Solomon is described in 1 Kings, the attention is given, in the words of one biblical scholar, to "his encyclopedic knowledge of the environment."[3] Solomon "spoke of trees, from the cedar that is in Lebanon to the hyssop that grows out of the wall; he spoke also of beasts, and of birds, and of reptiles, and of fish" (1 Kings 4:33).

The most powerful portrayal of God's relationship to creation within wisdom literature, and perhaps in all the Bible, is found at the end of the book of Job, in chapters 38 through 42. After Job's ordeal, he maintains, rightly, that his suffering is not punishment for sin. Still, he strives to know the reasons for God's actions; Job tries to rationalize God's actions into his own terms in order to gain some control and security in his situation.

In response to this almost arrogant striving to master the reasons for God's ways, the Lord answers Job out of the whirlwind. And God's answer is a profound description of God's intimate involvement in bringing forth and sustaining the elements and animals within the creation. The answer challenges Job with a series of questions asking, rhetorically, whether he possesses the power and knowledge necessary to sustain the creation. And even animals such as the "behemoth," or hippopotamus, and "leviathan," probably the crocodile, are described as beyond humanity's ability to control. This answer, of course, doesn't respond to Job on his own terms. Rather, God's role as Creator and sustainer is set forth as the basis for humanity's humble trust and confidence in the Lord.

These chapters give a marvelous and powerful picture of God's immanence, his dwelling within the creation. And they suggest an attitude of wonder toward the divinely governed ecological balance in the environment. In the face of these poetic descriptions of work in the created order, humanity's response should be one of awe, humility, and reverence.

God As Creator

For the people of Israel, faith in Yahweh began as a trust in God as the Creator. The world, and all that is in it, came forth and lives by the power of this one God. Therefore, the Israelites could place their confidence in Yahweh as the one to reign over nations, to deliver them from oppression, to bring judgment upon their foes, and to establish righteousness and peace. The foundation of this faith rested on God's action as the Creator. This was never seen as a mere footnote or explanation about how the world came to be. Rather, it was a dramatic act of trust, declared continually in the life of the people of Israel, which directed their worship and confirmed the path for faithfully following their Lord.

Psalm 96, mentioned earlier, gives an example of how Israel's faith in God as Creator provided a foundation for God's role as judge among the nations. The central declaration, "Say among the nations, 'The Lord reigns!'" is affirmed because this Lord is the Creator. "Yea, the world is established, it shall never be moved" (96:10). Likewise, Yahweh "is to be feared above all gods" and "all the gods of peoples are idols" because "the Lord made the heavens" (96:4, 5).

In the saga of the people of Israel, times of trouble and despair continually caused them to remember that their Lord is the Creator of all. This consistently became the basis for reaffirming their faith. During the dark days of exile in Babylon, for instance, the prophetic words of promise and assurance found in the second part of Isaiah point back to God's work as Creator. "'I am the Lord, who made all things, who stretched out the heavens alone, who spread out the earth . . . '" (Isa. 44:24; see also 40:12, 48:13).

For people of biblical faith, affirming God as Creator means asserting that the creation belongs to God. This is the starting point for a Christian's responsibility for the earth's environment. Contrary to our culture's commonplace assumptions, humanity cannot own the creation. "The earth is the Lord's and the fulness thereof," declares the psalmist (Ps. 24:1). And this simple truth echoes throughout the Bible's pages.

In Job, God's answer includes the declaration, "Whatever is under the whole heaven is mine" (Job 41:11). In Exodus 19, before the Ten Commandments are given, the Lord reminds Moses that "all the earth is mine" (19:5). In Deuteronomy 10, when Moses sets forth God's requirements to "fear the Lord your God, to walk in all his ways, to love him, to serve the Lord your God with all your heart and with all your soul, and to keep the commandments . . . " (10:12–13), he then immediately declares, "Behold, to the Lord your God belong heaven and the heaven of heavens, the earth with all that is in it" (10:14).

In Psalm 50:12, the Lord declares, "the world and all that is in it is mine." The New Testament repeats these statements. The prayer of praise at the end of Romans 11 concludes, "For from him and through him and to him are all things" (v. 36). And when Paul addresses the conflict in the Corinthian church over eating meat offered to idols, he quotes Psalm 24:1.

The conviction that the creation belongs to God, and not to humanity, was made concrete in the laws concerning the sabbatical and jubilee years for the people of Israel. "The land shall not be sold in perpetuity, for the land is mine," we read in Leviticus 25:23. The jubilee-year laws were intended to prevent the concentration of land possession, and insure its ongoing redistribution. The simple reason was that the land belonged to God. Likewise, the provision for the widows, orphans, and sojourners became a mark of Israel's faithfulness because these are the people who are landless. If the land truly is the Lord's, however, then all must share in reaping its fruit.

This definitive biblical declaration that the world belongs to God absolutely prevents any reading of Genesis 1:26–28 which would condone the selfish and destructive exploitation of the creation by humanity. The subduing and dominion of the environment described in these verses can be carried out only by affirming that all resources are God's. Human action must guard, preserve, and uphold God's purposes for the creation, as made clear in the instructions of Genesis 2:15 to tend and keep the garden.[4]

The Land Mourns

Our modern culture has all but forgotten this wisdom. A geography teacher at a state university in Washington sat next to me once on a flight to Seattle. After telling him briefly about my concerns and work, he looked down at the mountains in western Washington which were scarred by vast clear-cut areas—a forestry technique which cuts all the trees from a large parcel of land.

"The Scripture verse I'm reminded of is from Jeremiah," he suddenly announced, revealing a knowledge of the Bible. "It says, 'I brought you into a plentiful land to enjoy its fruits and its good things. But when you came in you defiled my land. . . . ' That's exactly what we've done."

Later I looked up the reference, and found it in Jeremiah 2:7–8. This is one of numerous biblical references portraying the unfaithfulness and sins of humanity expressed in the destruction of the environment. Yet, this relationship is more profound. Biblical passages frequently suggest that humanity's rebellion against God in any number of ways results in the land itself suffering, mourning, and becoming unfruitful.

The picture of the first sins in the Garden of Eden underscores the broken relationship which occurs not simply between humanity and God, but between humanity and the creation as well. Not only are Adam and Eve sent out from the Garden, which they wanted to treat as their own rather than obey God's limitations; they also encounter conflict rather than harmony in their relationship to the creation. And after Cain kills Abel, the ground itself cries out against the blood of this crime (Genesis 4:10). In his punishment, Cain becomes a wanderer, cut off from the fruitfulness of the earth.

This story is repeated in the experience of the people of Israel. In Isaiah 5 we read a condemnation of land-grabbing greed. "Woe to those who join house to house, who add field to field . . . " (5:8). What is the consequence? "For ten acres of vineyard shall yield but one bath [about six gallons], and a homer [about six bushels] of seed shall yield but an ephah [about one-half bushel]" (5:10). The sin of greed results in the unfruitfulness of the land. The same message is repeated by the prophet in even sterner terms in the twenty-fourth chapter:

> The earth mourns and withers,
>> the world languishes and withers;
>> the heavens languish together with the earth.
> The earth lies polluted
>> under its inhabitants;
> for they have transgressed the laws,
>> violated the statutes,
>> broken the everlasting covenant.
> Therefore a curse devours the earth,
>> and its inhabitants suffer for their guilt.
>> (Isa. 24:4–6)

In similar fashion, Hosea announces the consequences of unfaithfulness by the people of Israel. At the beginning of the fourth chapter, he describes the Lord's "controversy with the inhabitants of the land" (v. 1). Kindness and faith-fulness are lacking, replaced by lying, stealing, cheating, adultery, and murder. And what is the result?

> Therefore the land mourns,
>   and all who dwell in it languish,
> and also the beasts of the field,
>   and the birds of the air;
>   and even the fish of the sea are taken away.
>   (Hos. 4:3)

So the Bible clearly laments the deterioration of the environment. Contrasted with the wondrous pictures of creation's intended harmony and wholeness given in the Scriptures, environmental ruin is a direct offense against God the Creator. But more than this; biblical insight clearly names the cause of a degraded environment: human sin. Living according to selfish purposes, in alienation from God's purposes and love, quite literally causes the land to mourn, and the whole creation to be in travail.

Redeeming the Earth

Jeremiah asks plaintively, "How long will the land mourn, and the grass of every field wither?" (12:4). The biblical answer carries the promise for the renewal of the created order, continually springing fresh from the resources of God's grace. Just as God responds to human sin and rebellion with the invitation to new life, the response to the degradation of the earth is the concrete hope for restoring "shalom" and, in the words of the Psalm, the renewing of the face of the earth (104:30).

God's promises are framed through the covenant which the Lord establishes. The first mention of covenant in the Bible occurs immediately after the story of the flood in Genesis. When the creation, with all of its life, is reestablished as Noah and all the animals come forth from the ark, God's covenant is announced. And it is a covenant not merely between God and humanity, but rather with "every living creature," a covenant which God describes as "between me and the earth" (Gen. 9:13). To underscore the promise, the integrity of the earth's cycles—"seedtime and harvest, cold and heat, summer and winter, day and night"—is also assured (8:22), as the Lord promises, "I will never again curse the ground . . . " (8:21).

Since human sin has ecological consequences, so does the work of God's redemption. The biblical narrative continually sets forth the saving activity of God's grace, which not only delivers a people from oppression, but restores the life of all the creation. The goodness of the earth, and the environment's capacity to praise God's glory, are terribly marred by human rebellion. But, just as surely, it shares deeply in the redemptive work of God's grace.

Just as the first promise of covenant extends to all the creation, when Hosea describes God's steadfast love, affirming once again God's covenant, its scope remains just as expansive:

> I will make for you a covenant on that day with the beasts of the field, the birds of the air, and the creeping things on the ground; and I will abolish the bow, the sword, and war from the land; and I will make you lie down in safety. And I will betroth you to me for ever . . . and you shall know the Lord. And in that day, says the Lord, I will answer the heavens and they shall answer the earth; and the earth shall answer the grain, the wine, and the oil, and they shall answer Jezreel, and I will sow him for myself in the land.
>
> (Hos. 2:18–22)

In the same fashion, the redemptive promises set forth in the latter half of Isaiah are filled with images of all the creation participating in this saving and restoring work of God. "You shall no more be termed Forsaken, and your land shall no more be termed Desolate; but you shall be called My delight is in her, and your land Married" (62:4). The picture is of a redeemed people in a land embraced by God. "For you shall go out in joy, and be led forth in peace [shalom]; the mountains and hills before you shall break forth into singing, and all the trees of the field shall clap their hands" (55:12). So compelling is this promise that Isaiah even describes it as "new heavens and a new earth" (65:17). Redemption opens the world to a new creation.

Reconciling All Things in Christ

The New Testament builds on this foundation which integrates creation into the work of God's redemption. God's role as Creator and sustainer is ascribed to Jesus Christ in understanding the incarnation. As already mentioned, John says of Christ, "all things were made through him, and without him was not anything made that was made" (1:3). Colossians repeats this description with this worshipful declaration:

> He is the image of the invisible God, the first-born of all creation; for in him all things were created, in heaven and on earth . . . all things were created through him and for him. He is before all things, and in him all things hold together.
>
> (Col. 1:15–17)

The parables and teachings of Jesus are filled with examples drawn from the realm of nature. Vineyards, soil, fruit, seeds, and grain are the frequent examples used by Jesus to explain God's truth. And the Sermon on the Mount includes a direct, but often overlooked, teaching regarding our relationship to the creation. "Blessed are the meek," Jesus said, "for they shall inherit the earth" (Matt. 5:5).

Christ could well have had Psalm 37:11 in mind, which reads, " . . . the meek shall possess the land." His point, like that often heard in the Old Testament, is that the land, or the earth, can never be seized through our power and claimed as our own. Rather, it comes to us as a gift, as an inheritance from God, given only to those who do not grasp it, but rather care for it tenderly, meekly, and humbly.

When the work of God's redemption in Jesus Christ is discussed by New Testament writers, the reconciliation achieved through the life, death, and resurrection of Christ extends to the creation. Colossians, for example, declares, "For in him the fulness of God was pleased to dwell, and through him to reconcile to himself all things . . . " (1:19–20). This is the same "all things" which were created through Christ. Several other New Testament passages underscore how Christ's defeat of all the rebellious powers results in the restoration of God's purpose and intended order in all the creation.

God's love, then, extends to the entire world, to the whole creation. The Greek word used and translated as "world" is actually "cosmos." Often it refers to the whole creation. The same word also is used in the New Testament to mean that part of the world's life which is separated and alienated from God. When we read "be not of the world" or "love not the world," the meaning is to avoid the godless systems of life which break our relationship and trust in God. Further, we know that God's love reaches out to conquer the power of all that would separate us from the love of Christ. God's love for the world—for the whole cosmos—is the resounding biblical theme, and the reason for God's embrace of the world in Jesus Christ.

A close look at Paul's writing in Romans underscores these truths. At the beginning of the letter, he sets forth its theme in verses 16 and 17 of the first chapter. He describes the gospel as "the power of God for salvation," and then declares, "For in it the righteousness of God is revealed. . . ." What is meant by the "righteousness of God"?

Biblical scholars in recent decades have opened up rich insights concerning the full meaning of this term, as used here by Paul. The phrase has a rich history in the Old Testament, and is associated with God's saving activity toward the people of Israel. The righteousness of God indicates God's actions to save, to deliver, and to fulfill God's promises. Instead of meaning a legal status conferred on a person, the righteousness of God indicates the power of God's grace active in the world, bringing about salvation.

In Romans, then, Paul declares that in the gospel we

see fully God's power and righteousness displayed in the life, death, and resurrection of Jesus. God's purpose is to bring salvation to the world—a restored relationship with humanity and wholeness, or shalom, in the creation. And all this comes as God's free gift.

One perceptive biblical scholar, Ernst Käsemann, sums up Paul's meaning in this way:

> (the righteousness of God) is for Paul God's sovereignty over the world revealing itself eschatologically in Jesus. And, remembering the Greek root, we may also say that it is the rightful power with which God makes his cause to triumph in the world which has fallen away from him and which yet, as creation, is his inviolable possession.[5]

It is no wonder, then, that as Paul develops in Romans the theme of God's redemptive work, he comes to a climax exclaiming that the whole creation is eagerly longing to experience its full transformation. As one careful translation renders this powerful part of Romans:

> For the created order awaits, with eager longing, with neck outstretched, the full manifestation of the children of God. The futility or emptiness to which the created order is now subject is not something intrinsic to it. The Creator made the creation contingent, in his ordering, upon hope; for the creation itself has something to look forward to— namely, to be freed from its present enslavement to disintegration. The creation itself is to share in the freedom, in the glorious and undying goodness, of the children of God.[6]
>
> (Rom. 8:19–21)

From this and other passages, we can understand the relationship between God's work of redemption in our own lives and in all the creation. The final victory has been won by Christ. We belong to God. And the world belongs to God. At this moment, we who have encountered Jesus Christ know the first fruits of that redemption in our lives. We have become new, claimed by the power of God's Spirit. Likewise, the creation has entered into this renewal.

The power of sin in its midst, which has wrought destruction, is not the final word. Rather, the goodness of the creation, and its possession by God, is its final destiny.

Any counsel which suggests that Christians can simply ignore the desecration of the earth, believing it will be destroyed anyway, and that God only saves people's souls, flatly denies the truth of the Bible. Giving up the environment to the powers of destruction denies that the earth is the Lord's, and is in plain disobedience to the teaching of the biblical tradition which underlies Christian faith.

Creation gives God glory and honor. The gift of the environment came forth from God's will and power, and is to be a testimony to God's wonder and love. Christians have no less a calling than to participate in the preservation and renewal of this precious gift. With the words of Revelation, we can then proclaim in word and deed,

> "Worthy art thou, our Lord and God,
> to receive glory and honor and power,
> for thou didst create all things,
> and by thy will they existed and were created."
> (Rev. 4:11)

5

Hopeful Signs

Frank and Judy Alton are missionaries at a pastoral institute in Mexico. Their newsletter recently ended with this paragraph:

> We are learning over and over again here that the "earth is the Lord's and everything in it, the world, and all who live in it." The families with whom our students stayed did not want to be paid for this reason. What they have is for everyone. We have been without water off and on this summer. Our neighbors and friends, Irma and Fidencio, have a large cement hole beneath their yard which they fill when the water comes every 8 or 9 days. Because of this they do not run out so quickly. During the dry days we, and lots of other neighbors, carry our buckets to their house to fetch water. No one is refused, though they too could run out before the water comes again. These friends seem to understand that water belongs to everybody and everybody should have a little until it is gone. If the "earth is the Lord's and everything in it," then perhaps that would be true.

Earthkeeping is not a matter of biblical study and theology. Rather, it becomes a practical way of living. All over the world, Christians are beginning to apply the biblical call for preserving the earth to the practical circumstances of their own lives, their churches, and their communities.

My work with the New Creation Institute during the past few years has focused on writing, teaching, and promoting the church's responsibility to care for the creation. Five years ago, looking for Christians who were concerned about the environment seemed like searching for water in the Sahara. But today the situation is changing. Like water springing forth in the desert, an increasing number of Christian groups are making concrete expressions of earthkeeping. Sharing a few of these examples can be an inspiring way of answering the question, "What can I do?"

Recycle Unlimited

Calvin College in Grand Rapids, Michigan, is operated by the Christian Reformed Church. At first glance, this small, theologically conservative Christian college would seem like the most unlikely source of innovative environmental action. Yet, the Reformed theological tradition believes in building a thorough "world and life view"—in other words, formulating how Christian faith should influence each area of life in culture and society.

In the 1970s, a few students and faculty at Calvin decided that they should practice what their tradition preached, and figure out how to respond to the environmental crisis. So Recycle Unlimited was begun as an effort to use low, appropriate technology in the collection and processing of recyclable materials. A professor at Calvin, Dr. James Bosscher, encouraged students in the Engineering Club who designed an innovative recycling trailer that breaks down cans, plastic cartons, and bottles at pick-up locations. The concept was to encourage curbside pick-up and processing of household recyclable materials.

In 1976, experimental curbside collection of recyclable materials began in one neighborhood of Grand Rapids. By 1983, the City of Grand Rapids and the Department of Public Works in Kent County instituted an expanded citywide curbside service. The organizers of Recycle Unlimited calculated that 60 percent of the trash going to landfills consists of bottles, jars, glass, tin cans, plastic milk cartons, aluminum foil, and newspapers—all materials which could be recycled, saving 208 million tons of "trash" per year in Kent County.

The work, of course, has been slow, since it depends on the cooperation of people to separate the recycled materials and prepare them for curbside pick-up in Grand Rapids. Yet, in a recent year Recycle Unlimited had successfully recycled 7 million pounds of material. In about twelve years, it has become a credible, well-established, nonprofit alternative to the existing system of trash disposal. And the organization has received the endorsement of the area's major environmental organizations as well as respect and cooperation from county and city officials.

AuSable Trails Institute of Environmental Studies

In 1980, I received an invitation to a conference titled "The Environmental Crisis: The Ethical Dilemma." It was to be held at a place I had never heard of, the AuSable Trails Institute of Environmental Studies. I had moved from Washington, D.C. to Missoula, Montana earlier that year seeking a sabbatical time of reflection. And my concerns were beginning to focus on the relevance of Christian faith for environmental issues.

With curious anticipation, I made my way to the remote town of Mancelona, Michigan, near Traverse City, and arrived at AuSable Trails. For nearly all of the thirty participants, this forum was the first of its kind. Most were professors at small, evangelical Christian Colleges such as Westmont, Wheaton, Taylor, Gordon, Bethel, Dordt, and

Calvin. Several taught biology or other natural sciences. Everyone was gravely alarmed about the scope of the environmental crisis. And hardly anyone had ever gathered with a group of other Christians to discuss this issue.

This first AuSable Forum was typical of the catalytic role the AuSable Institute has come to play, particularly among Christian colleges, to stimulate the teaching and practice of caring for the earth. Located by a small, northern Michigan lake, the site originally served as a Christian youth camp. Then oil was discovered on the land. Suddenly the board, consisting of committed evangelicals, including some who taught at Taylor University, were confronted with perplexing choices. Should they pump out the oil? What would they do with anticipated revenues? What about Christian stewardship? Were there any environmental dangers?

Questions like those led to a dramatic decision in 1979 to transform the camp into a Christian environmental study center. The statement of purpose adopted by the board reads, in part, "The Board, faculty, and staff of the AuSable Institute confess that God is the owner of all. . . . The mission of the AuSable Institute is to be a center for study and experiences which integrate environmental information with Christian thought for the purpose of bringing the Christian community and the general public to a better understanding of the Creator and stewardship of his creation."

Pursuing this task, the AuSable Institute has focused on offering college courses to students who attend from other schools—usually for a quarter of study and field work. Working closely with the Christian College Coalition, AuSable attracts students from many evangelical and church-related colleges to offer special experiences in environmental education.

The forum in 1980 was the first in an annual series of events at AuSable bringing together theologians, teachers, environmental experts, and others for a week of discussion together. Year by year, topics have been wide ranging, including, "The Global Environmental Crisis and the Churches' Response," "Missionary Earthkeeping," and "Planotheonomics." The latter gathered Christian economists, ecologists, and theologians to discuss how economics and environmental concerns related together from a Christian point of view. An impressive series of papers as well as two books have emerged from these forums.[1]

Dr. Calvin DeWitt, mentioned in chapter 3, has given leadership to the AuSable Institute during these years. A graduate of Calvin College, DeWitt is a professor of environmental studies at the University of Wisconsin. His efforts have done much to open dialogue between the Christian community and those pursuing environmental studies in the academic community.

A recent study surveyed 125 church-related colleges to discover what courses they offered in environmental subjects.[2] Its encouraging finding was that 95 percent of these colleges are now offering course work in environmental subjects. In fact, when the survey was conducted, a total of 296 courses were available among these 125 church colleges.

Two decades ago, environmental studies were in their infancy, and "ecology" was an obscure academic term, unknown to the church. Today, Christian higher education has acknowledged the importance of teaching the care of God's creation. The present need is for courses that integrate biblical study, ethical considerations, and ecological issues. The AuSable Trails Institute of Environmental Studies can be expected to continue to play a critical role in stimulating the church and its academic institutions to participate in nurturing a sustainable earth.

Sierra Treks

Organized by Dave Willis, a graduate of Westmont College and Fuller Seminary, Sierra Treks is one example of a growing number of efforts to provide wilderness experiences for Christians that both deepen one's spiritual life and strengthen one's environmental commitments. For several years, Sierra Treks has been offering such wilderness opportunities to Christian groups. Some are offered for college credit through Westmont College.

While some trips have a focus on teaching wilderness skills and climbing, others offer serious study of forestry and wilderness issues in light of Christian theology. "Forest Ecology and Creation Theology," for instance, offered in 1986–87, included back-country trips in southwest Oregon for two weeks. It was available for four units of credit to Westmont students. The course description included these questions: "What's the difference between an old growth cathedral forest and a monocultural tree farm? What does such rapidly disappearing forest wilderness tell us about our theology and culture? And does the Bible have anything to say about relationships with the rest of creation?"

Other Sierra Treks activities work with churches, taking participants through the Glacier Peak Wilderness in Washington's North Cascades, and through the Hoover Wilderness in California's Sierra Mountains, as well as into Yosemite National Park, and many other locations. Willis himself has become deeply involved in political efforts to protect wilderness areas. And he offers a rich blend of theological insight and moving personal experiences to instill a love for the creation and the Creator in those who participate.

Throughout Christian tradition, the wilderness has been seen as a place for deepening one's utter dependence upon God. The wandering of the Israelites, the dwelling of John the Baptist, and the retreat by Jesus into the wilderness all underscore how God can encounter and empower people uniquely in this setting. The desert Fathers and many in the church's monastic movement followed this same call.

Our contemporary sense of wilderness may be more pristine and romantic than such biblical pictures. But those who fought in the early days of groups like the Sierra Club to preserve wilderness spoke passionately of its spiritual value. Sigurd Olson, a famous environmentalist, told a Sierra Club wilderness conference in 1961 this:

> I am happy to talk about the spiritual values of wilderness because I feel they are all-important—the real reason for all the practical things we must do to save wilderness. In the last analysis, it is the spiritual values we are really fighting to preserve.[3]

One contemporary author who captures well the capacity of wilderness for spiritual inspiration and nurture is David Douglas, a Presbyterian backpacker living in Santa Fe, New Mexico. His book *Wilderness Sojourn* gives testimony to the essential place of the wild for the life of the church and of culture today. In 1983, Douglas made similar arguments to the readers of the *Sierra Club* magazine, calling them to a spiritual and theological appreciation of wilderness:

> By dissolving, on a daily basis, the hubris that separated humans from God, the wilderness provided an arena of reconciliation. Wild country has always served to remind people of their limits, and it remains—in contrast to manicured parks, gardens and other cultivated landscapes—the one setting that a sojourner is unable to claim as his handiwork.[4]

Christians are deepening their appreciation of the wilderness as a place to be cherished and protected. The wild reminds us graphically that we are not the masters of the creation. In some settings, our call is to cultivate and tend, using creation's resources for harmony and fulfillment. In other settings, however, our task is to keep and preserve, so that the fullness of God's glory will have places where it is reflected in an unfettered way. Today, scores of people in the church are working, like Dave Willis and David Douglas, to link Christ's body with the spiritual value of wilderness.

Ecumenical Task Force of the
Niagara Frontier

In the spring of 1978, people living in a neighborhood on the east side of Niagara Falls, New York, noticed black, gooey matter seeping up to the surface in their backyards. At the schoolyard, rotting barrels, buried some time back, were uncovered. People noticed strong odors, and when it rained, pools of water colored from chemicals appeared. By August of that year, the New York Commissioner of Public Health issued a public emergency declaration and permanently evacuated 239 families whose homes bordered Love Canal.

Years earlier, wastes from the manufacture of pesticides, defoliants, disinfectants, plastics, and other products had been buried in this mile-long, abandoned canal. All had been forgotten until the spring of 1978. Unexplained illnesses, birth defects, miscarriages, and other health problems became linked to the poisonous earth as investigation revealed the toxic materials which had for years been infecting the families in that area.

After the initial evacuation and ongoing health problems and severe uncertainties, people began asking what the churches were doing to help. In response, the Ecumenical Task Force was formed to address the needs and the suffering caused by this manmade disaster. Its first gathering place was the Wesley United Methodist Church, which was later discovered to be resting on contaminated ground and was closed.

The Ecumenical Task Force faced escalating needs and immense challenges as the scope of the poisoning—and the tangled legal, political, technical, and psychological effects—became known. In addition to helping meet the needs of those victimized by this toxic disaster, the Task Force determined to educate the broader religious community concerning the dangers posed by poor storage of hazardous wastes. As the Task Force responded to this disaster, continuing to assist victims in the months and years that followed, other groups and individuals directed questions and concerns to them about possible dangers at other toxic waste sites.

The work of the Ecumenical Task Force of the Niagara Frontier has continued. Led by Sister Margeen Hoffmann, the organization has advised countless other groups and churches faced with similar hazards. In the decade since the Love Canal disaster first was revealed, the Ecumenical Task Force has served as a model for how churches can work together to prevent the further contamination of the earth.

Eco-Justice Working Group

Historically, the mainline Protestant denominations have championed causes of social justice as part of their witness within society. The civil rights movement, protests against the Vietnam War, and the antinuclear crusade all emerged from the early commitment of religious activists, and then built on the involvement of church people along with many others. But when Rachel Carson published *The Silent Spring,* which was catalytic in formulating environmental concern, and when Earth Day was launched in 1970, marking the political organization of the environmental movement, the church was absent.

Eventually, environmental questions began trickling into discussions within organized denominational structures. But almost always, these questions finally gained people's attention through other issues, such as energy policy, or hunger; no denomination had an office or staff person giving attention to the preservation of the earth. Social and economic justice concerns generally eclipsed the church's view of the environmental crisis. But in the late '70s, the Reverend John DeBoer led an interdenominational coordinating effort that became known as the "Eco-Justice Task Force." As its name suggests, this group attempted to bring together ecological concerns with the call to economic justice.

Providing resources for local churches, this task force launched two notable projects, one concerning energy conservation and the other concerning food stewardship. A helpful manual on how local congregations could save energy, and an eventual book on the use of solar-design architecture for church buildings were results of the first project. The organization's food stewardship efforts attempted to link those working on soil and farmland preservation, as well as organic farming techniques, with concerned church congregations in local areas throughout the country.

The National Council of Churches (NCC), meanwhile, became directly involved in the struggle over acid rain in 1984, including an important meeting with the United Church of Canada that considered how churches on both sides of the border could address this issue. The NCC's approach was not only that of looking for ways to halt the environmental damage caused by acid rain. It also considered the welfare of the Appalachian coal miners whose livelihood depended on mining the coal that, provided to midwestern power plants, was the cause of the acid rain. Environmental and economic-justice concerns were wedded together.

Later that year, the National Council of Churches drew the Eco-Justice Task Force under its umbrella, renaming it the Eco-Justice Working Group. This attempted to bring together in one place like-minded efforts primarily among mainline Protestant denominations. It continued its work on acid rain, including lobbying efforts on legislation in Washington, D.C.

The Eco-Justice Working Group also initiated dialogue with national environmental leaders and their organizations. Through this process, solid working relationships have begun to be established. And as the Working Group has turned its attention to specific issues such as groundwater contamination and toxic wastes, it is working in close coordination with environmental organizations devoted to addressing those problems. As a result, the efforts of both have been strengthened.

Today, environmental leaders are less apt to complain that the organized church has been inattentive to their struggle. For the first time since the environmental movement began, its leaders are finding themselves in meetings, dialogues, and coalitions with church representatives. And the interaction is healthy. As David Barker, director of Friends of the Earth, says:

> The churches need to tell environmentalists that ecological issues are social justice issues at heart. We need to talk about right and wrong. I am getting very tired of arguing that it is in the "self-interest" of people to work for a cleaner and safer environment.[5]

North American Conference on Christianity and Ecology

While some environmental initiatives within the church have involved primarily denominational staff people drafting statements and organizing lobbying activities, other efforts have sprung forth as genuine grass-roots movements of Christians wanting to preserve the earth. Such is the case with the North American Conference on Christianity and Ecology (NACCE). And surely this is an example of a tiny mustard seed growing into a broad tree.

In 1985, a half-dozen people began discussing the possibility of a major gathering in North America bringing together Christians concerned about the ecological crisis. None of them represented large church organizations. They were simply concerned individuals. Fred Krueger had been active in the 11th Commandment Fellowship, a loose collection of Christians with environmental concerns—many with some background in the Orthodox church. Father Al Fritsch had worked for years in Appalachia. David Haenke was involved in the bioregional political movement, an initiative related to the U.S. "greens," activists exploring political options similar to the Green Parties in European politics. Thomas Berry, mentioned in chapter 4, also joined with them, along with Cal DeWitt.

In those early days, potentially interested groups and individuals were contacted about the vision of a national conference. I first came in touch with them, for example, when they were just forming a steering committee and attempting to find out who else might share such a vision. What impressed me from the start was that this initiative was entirely outside any existing organizational structure, and that it was committed to being wholly and actively ecumenical. All Christians who shared these concerns were to be brought together.

Its first statement of purpose announced that its goals would be "to elucidate Christianity's inherent ecological dimension. . . . To transform every church into a citadel of ecological understanding and action, and to make every Christian an ecologist." It is no wonder that some observers viewed the whole initiative with measured skepticism. And yet, the organizers persevered, meeting with interested supporters to form a board fifteen months before the conference was to take place.

The enormous theological diversity of NACCE, as described briefly in chapter 4, has been both its unique characteristic as well as its greatest source of stress. Its statement of purpose also said, "we seek a contemporary exposition of the eternal and unchanging truths ever present in Christianity." From the beginning, however, there were tensions between some versions of "contemporary exposition" and other views of "unchanging truths."

Nevertheless, by the time of the actual conference in August 1987, the vision of NACCE's organizers had been rewarded with a groundswell of response and participation. Sixty-three groups and organizations had become sponsors of the event by that time, and five hundred people attended, coming from virtually every state in the union, as well as Canada, Mexico, and Europe.

Held at the United Methodist Church's Epworth

Forest Conference Center in Indiana, the conference included sixty workshops covering everything from the Eastern Christian view of ecology to the crisis of the family farm. Participants seemed genuinely surprised, as well as encouraged, to discover the breadth of individuals and groups which responded to the event. As one person said, "The Christian churches have belatedly entered the environmental movement."

At the conference, Board Chair Al Fritsch stated, "Our goal is a practical one. It's to serve those called to lead a greater environmental awareness within the churches." NACCE has continued that goal since its conference. With an ongoing office in San Francisco, it seeks to act as a clearing house for information and activities expressing the church's growing ecological concern. Other regional gatherings are also planned.

Perhaps more than any other recent event, the North American Conference on Christianity and Ecology demonstrated to the church as well as to the public that growing numbers of Christians believe that their faith compels them to defend the gift of God's creation. And it also has given witness to the genuine, grass-roots nature of this movement.

Christian Farmers Federations in Canada

Preserving the creation is not an academic issue on the farmlands of North America. There, the severe immediate economic pressures facing farmers often seem to collide with the mandate to "tend the garden" in order to insure the fruitfulness of its soil for future generations. Ultimately, good economics and good ecology become the same. And some Christian initiatives in rural areas are attempting to demonstrate this truth.

Among the more unique, and unknown, examples are the Christian Farmers Federations in Ontario and Alberta, Canada. Begun largely by farmers from the Christian Reformed Churches in Canada, these groups attempt to explore practical ways to live out their stewardship of the land as farmers. The two federations also publish a quarterly magazine, *Earthkeeping,* devoted to faith and agriculture.

An editorial from a 1987 issue of the magazine begins with this explanation:

> A good farmer has insight into the creational wisdom of God and, in his or her work, touches on God's guidelines for agriculture. This awareness enables a farmer to shape a practice of working the land in a way that keeps and perhaps improves the earth.

Concern with the just, stewardly provision of a basic creational need—food—motivates Christian Farmers Federation members to consciously care for the land in a world which belongs to God.[6]

Federation members have also traveled to countries such as Mexico to understand the struggles of poor Mexican farmers. And they have studied problems such as multinational companies utilizing vast lands to grow food for export while the rural population suffers from poverty and inadequate nutrition.

These Christian Farmers Federation members still have their arguments with environmentalists over various issues. Yet, they are earnestly seeking to live out their vocation with a sense of serious responsibility to God and toward the care of God's creation. Their thoughtful efforts are instructive for anyone struggling to understand the task of earthkeeping in the midst of an agricultural system that allows profit to erode stewardship.

Within the United States, many organizations prompted by the church have become involved in efforts to conserve eroding soil, protect smaller, family farms, and build a decentralized, environmentally sound food system. The key component is finding ways to enable more food to be grown and consumed locally. This can encourage organic farming practices and can begin to eliminate the enormous energy and packing costs which presently dominate our food system.

One simple example is illustrative. The Land Stewardship Project is located in Winona County, Minnesota. Its many activities work throughout the upper Midwest to promote sustainable agriculture. But in Winona, it supported the Redeemer Land Stewardship Garden. The Redeemer Lutheran Church donated two acres of land. Volunteers along with developmentally disabled adults worked together on the land. A four-hundred-tree windbreak was planted, both to reduce erosion and enhance beauty. And all organic farming methods were used. The result was the production of seventy-five hundred pounds of organic produce in two years' time for the shelves of local food stores.

Missionary Earthkeeping

"Poverty is an ecological matter. The left asks why there is so much poverty in a rich country, but the Philippines is no longer a rich country, not after years of agribusiness and logging. Until there is a comprehensive understanding of how things stay alive in the Philippines, there will be no comprehensive understanding of what it takes to make peace."

So said Columbian Mission Father Vincent Busch at the North American Conference on Christianity and Ecology. He echoes the experience of hundreds of missionaries and church workers throughout the impoverished countries of the world. Increasingly, the links between ecological balance and economic justice are becoming clear in those areas.

I first saw this most graphically in 1983 at a conference in India, cosponsored by the New Creation Institute. Titled "Christian Perspectives on Stewardship of the Earth's Resources," this first-of-its-kind gathering brought together rural development workers, Christian theologians, leaders in the Indian church, community health workers, and others, who gathered at the Christian Medical College in Vellore. As we listened to the description of falling water tables, erosion, and deforestation, the devastating impact on the rural poor became immediately clear.

The church in India, for the most part, knows that the gospel must address the needs of the whole person. That probably explains why the Christians at this conference were quick to understand that preserving the environment, working for economic justice, furthering the ministry of healing, and proclaiming the gospel all fit harmoniously together as the church's call and witness within that society. Those attending the conference agreed on a statement which addressed their concerns, in part, this way:

> We have become deeply alarmed to learn, during our Symposium, of the grave dangers posed to the environment in our country. Available water is being rapidly depleted by agricultural practices which squander this resource. . . . Forestry practices have resulted in lands being denuded of trees and vegetation for short-term commercial gain, while worsening the problems of water scarcity and erosion. . . . Such inter-related problems increase the suffering of the weaker sections (the poor) in our society, who experience the effects most immediately, and threaten the ability of the Creator's resources to nourish life for now and the future. . . .
>
> The biblical imperatives for justice, and its call to care for the creation, combine together. The church's ministry of healing and wholeness calls us to the work of earthkeeping as we proclaim in word and deed the gospel of Jesus Christ.

The same integrated and wholistic expression of the gospel was shown to me in the work of Jim and Joan Gustafson, missionaries of the Evangelical Covenant Church in northeast Thailand. There, an innovative approach links the establishment of new churches and discipleship with the forming of cooperatives and the development of a sustainable, environmentally sound model of pig and fish production. Those in the church have the opportunity to form a cooperative which begins with pig breeding. Then a pond is dug and fish are added, using pig manure as rich nutrients in the water to provide their food. Banana and other fruit trees are then planted on the dikes surrounding the pond, and ducks are added.

This amazing model has enjoyed success in beginning scores of new churches, teaching new converts the meaning of discipleship in very concrete ways—how to live cooperatively rather than selfishly, for example—and then providing the means for ecologically sound, sustainable economic development. All this is seen as an integrated whole. People do not deal with evangelism on one hand, and then various social "issues" on the other. Rather, all of life, related to the creation, is encountered by the full message of God's grace in Jesus Christ.

Examples like these can be cited from throughout the Third World. The church's mission in the world includes earthkeeping.

Justice, Peace, and Integrity of Creation

When the World Council of Churches held its Sixth Assembly during 1983 in Vancouver, Canada, it searched for the commitments that could guide its program outreach into the next decade. The theme which the WCC adopted was "Justice, Peace, and Integrity of Creation."

Most would expect the WCC to be active in global issues concerning justice and peace. But the inclusion of "integrity of creation" made a new statement concerning the WCC's involvement in global concerns. These leaders of the world church were recognizing, almost prophetically, that any program of mission addressing the world's critical problems must confront directly the threats to the integrity of the creation.

Since then, the World Council of Churches has served as catalyst to many parts of the global Christian community, raising the challenge of protecting God's creation. Its

Faith and Order Commission met in the late fall of 1987 to consider "The Doctrine of Creation and Its Integrity." Early 1988 witnessed a WCC-sponsored international consultation on the integrity of creation, held in Norway. And a world convocation on "Justice, Peace, and Integrity of Creation" is planned for 1990.

The World Council of Churches' attentiveness to this subject indicates the growing sensitivity of the church to the globe's environmental crisis. As one surveys the number of emerging groups and new efforts working today for environmental preservation, it seems as though a fresh movement of the Spirit can be sensed, seeking to renew the earth once again. The church is being called to the forgotten task of saving the earth. Christians everywhere who hear this call can now find specific, practical avenues for living it out.

6

Cooperative Technology

One of my simple joys is working with the compost bins in my backyard. A few years ago, a friend and I built them using a design from a book titled *The Integral Urban House*. Throughout the year, I dump grass clippings, weeds, plants that have born their fruit, flower stalks, kitchen wastes, spoiled produce, torn-up sod, ashes, leaves, and, when I'm lucky, rabbit manure, into the bins. Periodically, I turn and mix the materials, sometimes adding needed ingredients.

Then, usually in the fall and the spring, I take the compost which has become well broken-down, and dig it into the garden and flower beds. Doing this always gives me a wonderfully satisfied and rooted feeling. There is something special about strawberries, tomatoes, and flowers that grow from soil made rich from "wastes" which have become nutrients.

Science and technology help make this process possible. Composting, of course, is not what normally comes to mind when one thinks about scientific and technological achievements. Yet, the composting I do depends on such knowledge. In fact, if I had more scientific knowledge and better technological tools, the composting I do could be vastly improved.

Composting actually involves a lot of science. It all began through scientific observation which discovered bacteria and then observed how they function within organic material to break it down, in the process of decay. As microorganisms do their job, taking nitrogen from organic materials and then dying, the nutrients in their bodies can become available to plants. And as these soluble nutrients become lodged in humus, they are ideally suited to nourish plants over an extended period.

For this to happen in aerobic composting, as opposed to anaerobic, there needs to be the proper mix of carbon to nitrogen—about twenty-five to one. The different materials put into a compost pile are composed of varying proportions of carbon and nitrogen; so sorting them out also depends, of course, on scientific knowledge. Then, a technological method is required to make this all work. The first prerequisite is building the proper kind of container, which depends on one's composting method. And there are various formulas which can be used for applying different materials in layers to start the compost pile.

In aerobic composting, which refers simply to composting that depends on oxygen being fed to bacteria and fungi, the technological methods result in temperatures within the pile of up to 160 degrees, if all goes well. And the result, under the best of circumstances, is dark, rich humus full of the nutrients needed by plants to grow and flourish.

So composting involves science and technology. Yet, it doesn't try to conquer nature. Rather, science and technology are used to enable an innovative means for cooperating with a natural process. We draw upon the resources and capabilities within the creation in ways that utilize its self-sustaining potential, in order to provide us with food, flowers, and plants. Our stance toward the creation in the process is more like participation than domination.

Life as we know it would be impossible without science and technology. Upon reflection, no one would seriously suggest that we should, or even could, discard certain aspects of life in favor of a totally "natural," pristine existence. As the example of composting makes clear, activities that are sometimes described as "natural" methods do not reject science and technology, but actually depend on them. Yet, there is a clear difference between applying compost from my bin to my tomato plants, and pouring on a bag of chemical fertilizers derived from petroleum and other products.

A Christian commitment to care for the gifts of God's creation ultimately gets translated into values and concrete choices that are made regarding the place of technology. Clarifying such values and making those choices can be assisted by our gaining perspective on our culture's assumptions regarding science and technology.

Beyond doubt, technology has changed everything in modern life. Our relationship to time, our goals, the way in which we think, the pattern of our relationships with others, and our ideas of pleasure and satisfaction have all been vastly molded by the development of technology. By and large, this is merely accepted—if even recognized. Rarely do we realize the extent to which modern technology, in and of itself, sets forth the terms, vision, and values for our lives.

The power of technology within our culture has resulted from the scientific revolution. Prior to that time, the pursuit of metaphysics (reflection on the nature of reality) and applied technology (how to get things done) were not closely related. But with the scientific revolution came a presumed certainty in the knowledge of the world. And the purpose of this scientific knowledge, as explained carefully by Francis Bacon, was to gain power and control over nature. Science would supply the knowledge necessary for technology to exploit its rewards from the natural world. Science and technology became linked together.

These forces ushered in the industrial revolution and reshaped the fabric of modern life. In the process, society came to assume that the course of progress, human betterment, and social well-being would unfold naturally from the development of technology. This was the means of humanity's liberation. No room was left to question the influence and effects technology was having upon cultural goals and values. Thus, we arrived at our present predicament—technology has become an idol, an object of faith, and an oppressive power rather than a servant of human fulfillment and the wholeness of creation.

The dangers of this reign of technology's power can be summarized in four ways.

First, modern society has come to place a fundamental faith in technology itself. An intrinsic belief assumes that problems faced by the society will yield to technological solutions. Whether issues of hunger, poverty, energy use, environmental deterioration, disease, the AIDS epidemic, or depletion of the ozone layer, a confident faith insists that *all* these will be solved by scientific and technological breakthroughs.

Even when we recognize that many of these problems have, in fact, resulted from the impact of technology, there remains the clear expectation that technology can make right what has gone wrong. In many cases, technological achievements have brought solutions to major problems facing society. This we can certainly affirm. But that does not justify the blind faith which society places in the technological process. The assumption that all problems will yield to technological solutions prevents us from considering how the following could be essential requirements:

- changes in basic values
- adjustments in lifestyle
- expansion of social compassion and sacrificial service within society
- the power of moral and ethical commitments

- modification of society's basic goals
- acceptance of political and economic changes

Second, society assumes that technological knowledge is omniscient. Whatever society needs to know about a given issue involving technological choices, technology itself will provide. Thus, the careful analysis of cost-benefit ratios, computerized calculations of acceptable risk, and studious assessments of environmental impact provide the necessary "inputs" for making choices and decisions. Such procedures certainly have a place. But can they fully replace the function of moral judgment, ethical intuition, and spiritual vision? Can a society survive if it pretends that such "unquantifiable" factors cannot influence decisions about clearcutting a forest or building a power plant?

Third, technology has imposed its own rules as society's features. As technology has developed within our society, it demonstrates two chief characteristics—growth and efficiency. Those qualities, in turn, have become embraced by society as a whole. In most all areas of life, we assume that bigger means better, and that speed and efficiency are almost always to be highly desired.

Fourth, technological ends have become translated into society's goals. Technology is driven by the need to increase our comfort, pleasure, entertainment, and security; few would wish for all those benefits to be denied. However, the intensity of technological development tends to imprint those benefits as the chief goals of the society. They become the ends of life itself. Social, political, and economic forces are then geared to insuring human comfort, "happiness," and security. Moreover, these are defined in purely material, quantifiable ways—the "standard" of living.

Challenging Technological Idolatry

Faced with these consequences of technological power, Christian faith must challenge such idolatry with the vision of wholeness for the creation and true fulfillment in human life. Society's faith in the utter resourcefulness of technology, its belief in the sufficiency of technological knowledge, its absorption of technology's rules, and its embrace of technological ends all should be countered with warnings and with deeds which regard technology as our servant rather than our master.

Such hopes are not naive. Ironically, the society which has become the most technologically advanced may also be the most open to sensing the limitations of technology. As Albert Borgmann has written:

> From a look at the ways in which technology has infected and transformed all cultures of the globe it appears that there is hope for a coming to terms with technology not in the vortex of the initial confrontation, but only after one has passed through it.[1]

Modern society displays many signs which evidence a recognition of technology's limitations and a search for values that redress technological dominance. In agriculture, organic farming techniques are finding a growth in support and acceptance. Agricultural practices that attempt to cherish the land, and draw upon the intrinsic strengths of the created order, rather than impose systems of agribusiness production, are gaining increased notice. The work of Wes Jackson at the Land Institute in Salina, Kansas is most noteworthy in this regard.

The dominance of technology in medical care is also being reconsidered. In fact, what amounts to virtually an alternative system of health care—emphasizing an eclectic mix of natural remedies, wholistic practices, the hospice movement, and wellness programs—is challenging a traditionally blind faith in medical technology. In as basic an experience as childbirth, a resurgence of interest in "natural" methods today is contrasted to, and often blended with, procedures that previously treated childbirth as a medical emergency demanding technological mastery.

A theology of creation directs the church into those movements that are reevaluating the role of technology within the society. From a biblical view, we are warned against the mistaken urge to impose a rigid technological control and mastery over the created order. As resources are exhausted and the environment is pushed to the breaking point, the biblical imperative of caring for, preserving, and nurturing the created order as God's gift becomes essential. And technology then can become a tool and servant safeguarding the integrity of the creation.

When such broader social values, moral wisdom, and spiritual vision provide the framework in which technology can function, then the relationship between technology and the creation is transformed. Such new expressions of technology demonstrate innovative means for cooperating with the created order rather than conquering it. This is obvious in organic farming, solar energy, and natural childbirth. "Appropriate technology" also expresses this relationship as it has reshaped the place of technology in rural development. Moreover, even the more highly complex forms of technology, such as computer analysis of satellite photography, can serve the same end of assisting our participation with the resources of the creation, rather than hastening their depletion.

Technological progress now presents creation with the threat of destruction, and confronts the church with a radical theological and practical challenge. Yet, the biblical vision of shalom remains the hope for the world's salvation. By living out of this vision, the church can demonstrate how the relationship between technology and creation can be redeemed. Turning from the idolatry of technology, we can help fashion forms of technological innovation that put us back in touch with the wisdom of God as reflected in the wonder of creation's gifts, and their capacity for sustaining life.

Engineering Life

Few issues pose greater challenges to the church's need to evaluate technology than the advances of genetic engineering. Just as splitting the atom granted to humanity unprecedented power of destruction, splicing genes has conferred on us drastic powers to create new life forms and alter the basic features and boundaries of God's creation.

The potential consequences are vast. The genetic features of humanity itself are subject to alteration through this technology. Many religious leaders have issued strong warnings against techniques that would produce changes in the human germ line. Permanently correcting genetic "flaws" assumes that there exists an ideal of human perfection known to us. Since qualities such as skin color, sex, and height, as well as various mental and physical abilities, could be structured by the techniques of human genetic engineering, the prospect of creating "superior" people through eugenics becomes a chilling possibility.

A line must be drawn between the curing of hereditary diseases and illnesses, and the reconstitution of humanity's genetic characteristics. The persistent concern in the Scriptures for healing and restoration, and the healing ministry of Christ, point us toward God's intended purposes for humanity and all creation. Yet, that same concern calls us to a stance of treasuring life in all its expressions. The potential in human genetic engineering is to slip very easily from healing and restoring life to the redesign of basic human characteristics. This danger cannot be underestimated, for the technological possibilities are within humanity's grasp.[2]

On the other hand, techniques of gene therapy on an embryo (somatic-cell therapy) are being explored which could offer cures to certain genetic diseases within that individual, without affecting the human germ line. It is possible that such measures can offer a cure to certain hereditary diseases that otherwise cannot be treated.

From the standpoint of our theology of creation and the biblical responsibility for preserving the environment, the application of genetic engineering to plants and animals raises critical challenges. I have already referred to the 1980 action by the United States Supreme Court which ruled that a genetically engineered bacterium, a new life form, could be granted a patent. The ruling suggested, philosophically, that "life is largely chemistry," and conferred the rights of ownership and patenting on these newly created microorganisms.

Critics warned that this would set the stage for the development and patenting of new life forms in the animal world, eventually even combining human genetic characteristics with those of animals. At the time, many dismissed such predictions as improbable. But by 1987 they had become a reality. The U.S. Patent Office in that year issued a ruling, based on the 1980 Supreme Court decision, that genetically altered animals, combining traits of different animal species as well as human genetic characteristics, could be granted patents.

The ruling shocked many environmentalists, scientists, public-interest groups, religious leaders, and politicians. Senator Mark O. Hatfield introduced legislation requiring a moratorium on that action. He declared on the Senate floor, "The patenting of animals brings up the central ethical issue of reverence for life. Will future generations follow the ethic of this patent policy and view life as mere chemical manufacture and invention, with no greater value or meaning than industrial products?" His measure passed the Senate, but action in the House of Representatives was inconclusive, resulting in congressional hearings on the subject. Meanwhile, the Patent Office delayed implementation of its ruling into 1988, partially waiting to see if Congress would take decisive action. Then, in April of that year, it issued its first patent to such an animal, a mouse altered and developed for medical research and owned by the DuPont Company.

The development and patenting of transgenetic animals is an unprecedented shift in humanity's relationship to the God-given natural environment. These actions raise critical questions for the religious community. The church should be alarmed that technological achievements in biotechnology proceed with breath-taking speed, while moral and ethical reflection has barely been heard. Most alarming is the fact that the pressure to develop and patent genetically altered animals would allow the marketplace to override any dictates of morality and religious values, and regard new animal-life forms solely as matters of economic utility.

For this reason, the National Council of Churches urged Congress to adopt measures halting the implementation of the April 7, 1987 ruling by the U.S. Patent Office. As stated by the general secretary, the Reverend Arie Brouwer:

> No decisions about the patenting of genetically altered animals should be made without the most careful examination of all possible consequences. Our culture's stance toward the gifts of God's creation, and our respect for all life, call for thoughtful reflection and judgement on these matters by churches and religious institutions, as well as by other concerned groups in our society.

Therefore, the National Council of Churches supported "congressional measures to halt such actions by the Patent Office and to enable a careful consideration of these questions by Congress and the public." That statement was agreed to by the leaders of most every major Protestant denomination, including the United Methodist Church, the Presbyterian Church in the U.S., the Episcopal Church, the Lutheran Church in America, the United Church of Christ, the American Baptist Church, the Reformed Church in America, the Disciples of Christ, and others, as well as representatives from the Orthodox Church. Additionally, leaders of major Jewish groups joined in issuing this plea.

In short, the prospect of patenting genetically altered animals with little or no public notice and discussion provoked an alarmed reaction from America's religious community.

The National Council of Churches and other religious leaders did not urge such action because of any blind opposition to genetic engineering and biotechnology. On the contrary, the NCC's official policy statement adopted in May 1986, "Genetic Science for Human Benefit," represented a careful attempt both to recognize the potential benefits of this technology as well as identify the crucial ethical and moral questions which it raises.

The problem, however, is that the rapid pace of this technology has been outstripping society's capacity for any considered moral judgment. That is precisely the situation presented by the Patent Office's ruling permitting the patenting of genetically engineered animals.

It may be the case, for instance, that extracting the human growth hormone from a cadaver and inserting it into a pig can result in what one scientist, Dr. Thomas Wagner, called "fundamental increased molecular efficiencies,"[3] decreasing feed costs and increasing someone's profits. But does this mean that it is right?

Are we to assume that nothing should prevent us from extracting genetic characteristics from a human being and inserting them into animals? And do we not have qualms about regarding such an altered animal an "invention" owned and patented by a commercial company? Should not the possibility of combining human and animal genetic material into novel technological creations compel rigorous moral, ethical, and religious consideration?

We have seen, in chapter 4, how the Bible emphasizes the goodness of the created order, and pictures the creation as giving praise to God. Moreover, God is portrayed as sustaining the wondrous, ongoing gift of creation's life. Humanity's unique power to discover, shape, and utilize creation's resources is to be guided by the need to uphold this life-giving environment.

Such insights form the basis for attitudes of respect and reverence for the nonhuman elements of the created world. This stance has been transmitted into some of our cultural values, as well as laws that protect endangered species, promote animal welfare, and safeguard environmental quality.

On this basis, Christians should seek to restrain those actions which threaten to undo basic features of the created order. The production of new animal-life forms which combine the genetic qualities of different species is such an example.

Observing recent innovations of biotechnology—such as combining genetic characteristics of cows with pigs; inserting bovine growth material into salmon, creating superfish; and uniting the phosphorescence of fireflies with tobacco plants—one gets the clear feeling that creation's inherent structures and boundaries are of little intrinsic worth. The stance assumed by such actions is that God's creation is lacking in sufficient wisdom, and fully in need of being fundamentally restructured, for the sake of furthering profits.

The patenting of such genetically engineered life forms, furthermore, makes an economic assumption that these creations will serve the public interest. It answers any questions concerning their value in solely economic terms, and without debate. On this basis, a corporation's actual ownership of a new form of animal life, with all its offspring, is recognized and protected.

Those developments caused Rev. Arie Brouwer and several other leaders, echoing the sentiments of Senator Hatfield, to warn, "Reverence for all life created by God may be eroded by subtle economic pressures to view animal life as if it were an industrial product invented and manufactured by humans."

Arguments that altering animals through genetic engineering is much the same as the selected breeding of animal species seriously miss the point. Natural breeding to improve an animal species exercises our human responsibility to wisely nurture and utilize creation's resources. But the engineering of transgenetic animals produces life forms which are impossible to achieve through any natural breeding process. Humans don't breed with rabbits, nor cows with pigs, much less fireflies with tobacco plants.

Some have argued that the existing examples of transgenetic animals, such as pigs which mature faster, or livestock with human blood-clotting genes, seem to present no harm, and serve some economic purpose. But where are the limitations on such activities? What will prevent any possible new life form from being created? Who will decide this?

One could think of this analogy. Suppose an owner of a vast estate placed his home and property in trust to a caretaker for many years. The caretaker's responsibility would be to preserve and keep up the property to the full satisfaction of the owner. The house would need painting and repair. Floors would be polished, windows replaced, and plumbing repaired. Further, the grounds and gardens would require constant care.

But what if the caretaker, who was neither the architect or builder, decided to remodel the house? What if he didn't like the kitchen, and tore it out, replacing it with another design? What if he took out the walls between three bedrooms to make a recreation room, destroyed the porch, and tried to add a swimming pool? And suppose he bulldozed down an orchard, wanting to create a lagoon?

First, he would have to possess the skill and knowledge necessary to accomplish this without damaging the house and grounds. But second, he would be arrogantly assuming the role not of caretaker, but rather master builder and owner. And this would violate the entire concept of placing a property in trust to another. Holding something in trust imposes both responsibilities and limitations. The technological possibilities of genetic engineering threaten to evade this entire understanding.

Patenting new animal life forms is like crossing the Rubicon. It is a decision with potentially momentous consequences, not easily undone. Christians must strongly counsel caution and restraint in all those areas of biotechnology which pose unanswered moral and ethical questions and assume unprecedented changes in humanity's intervention in the created order.

Certainly, the burden of proof should rest with those who advise speed and quick action on genetic engineering's innovations such as animal patenting, rather than on those who counsel caution, who desire a full public dialogue, and who propose careful ethical and theological reflection.

We should remember that all innovation is not progress. And hasty interventions in the environment have often wrought unforeseen and costly results. The gypsy moth, the kudzu vine, detergents, and DDT are but a few examples.

We have good reason to doubt that humanity knows enough to restructure the basic forms of life in the animal kingdom. Producing such new life forms simply for the sake of profit is morally offensive and wholly unwarranted.

The advent of biotechnology and genetic engineering is presenting society with fascinating and intriguing tools. Like other technological innovations, they present promises as well as dangers. But, as one church policy statement put it, "The sudden leap from laboratory to factory has seldom, if ever, been made so rapidly as in the case of genetic research."[4]

Faced with these new and powerful tools, society and the church must determine what we will build with them, for whom, and why. And to answer such questions, we need wise people, not just clever people.

The Psalms remind us that "the earth is the Lord's, and the fullness thereof." Before deciding the questions of the patenting and owning of new animal life forms, the church must insist that our society ponder the responsibilities entrusted to it for safeguarding the gift of creation's resources. For in the end, we only hold creation in trust. Our use of any technology must always be judged according to the standard of this biblical wisdom.

7

This Creation and the

New Creation

The promise of a new creation is central to Christian faith. In Jesus Christ, humanity witnesses human life being made new. The message of the gospel, in part, is that life for any person can be radically changed; instead of living in bondage to self-serving and ultimately destructive forces, one's life can be transformed by the same Spirit which was fully present in Jesus. The previous reign of sin in one's life is overcome as one, united with Christ, experiences forgiveness and becomes a new person.

Explanations of this experience often quote Paul saying that "the old has passed away, and the new has come," from 2 Corinthians 5:17. Those circles within the church emphasizing the centrality of the conversion experience maintain a sharp distinction, or even barrier, between one's old life and one's new life. Being "born again" means that one's life before this spiritual rebirth is, in fact, dead, and that an entirely new person has now been created.

Paul's words in 2 Corinthians 5, however, point to this new creation as involving far more than just the person. In fact, a strong exegetical case can be made that this passage on new creation concerns not the promise of internal new life to the individual at all, but rather, newness for the entire creation. "If any one is in Christ," Paul begins in verse 17; immediately we are reminded of the meaning of the phrase "in Christ," used so often by Paul and explored thoroughly by New Testament scholarship. Rather than the inner experience of having Jesus in one's heart, Paul means the individual's belonging to the body of Christ, the church—this new group of people bound to each other and God through their common faith in Jesus as Christ and Lord, and expressing in their corporate life the signs of this new reality in the world.

The Greek text continues with the words "new creation." (The English phrases "there is" or "he/she is" do not commonly appear in Greek, as these are assumed through grammar and sentence structure.) Thus, Paul maintains that when one enters into the body of Christ, the "new creation" is present to that person. Rather than assuming this new creation applies only to the individual, and translating the passage "he is a new creation," as in the RSV, a more appropriate rendering would be "there is a new creation," or, as in the New English Bible, "When anyone is united to Christ, there is a new world; the old order has gone, and a new order has already begun."

Thus, the biblical promise of "new creation" means just what it says—a new world, a new created order. Constricting this promise to just the individual's experience of rebirth misunderstands the biblical message. The first appearance of this hope comes in God's promises to the people of Israel. Even the first eleven chapters of Genesis tell the story of the original creation almost erased by a deluge, and yielding after the flood to a new creation, explicitly embraced by the Noahic covenant announced in the ninth chapter.

Then, the promises of liberation from oppression and

inheritance of a new land express the hope of a new society. This new society will live in accord with a right relationship to God and to the earth, all of which would fulfill the vision of shalom—wholeness, fruitfulness, fellowship, and peace. Later, the exilic hopes of Israel looked to a new messianic age described in Isaiah 65 as the creation of "new heavens and a new earth."

Yet, a counterpoint, often neglected by Western theology, runs with equal strength through the Old Testament's pages. God's grace and fidelity are discovered within the given creation. Even the Covenant with all creation declared by God after the flood underscores the trustworthiness of the created order, in spite of humanity's disobedience. "While the earth remains, seedtime and harvest, cold and heat, summer and winter, day and night, shall not cease" (Gen. 8:22).

As discussed in chapter 4, the wisdom literature in the Old Testament richly affirms the discovery of God's order and truth through humanity's experiential relationship with the creation. The earth is good, and God's ways become known as humanity is receptive and open to this search. We are neither liberated from the present creation, nor do we harshly impose a human order upon it. Rather, we learn to observe God's orderliness within the creation, and then live accordingly.

The Old Testament, then, presents God's activity as embracing all the creation. At some points, the stress is placed on prophetic judgment of present realities and hope for a new order, often remembering the reign of David and anticipating a similar but more universal glory in the future. At other points, the task becomes discerning the reality of God's presence and truth within the given circumstances of life.

Thus, the biblical promise of the new creation begins not with the person as the focus, but with the world. The New Testament carries forward and culminates this perspective. Much of our theology, however, under the influence of the entrenched individualism of modern Western culture, has assumed the opposite. From Billy Graham to Rudolf Bultmann, the focus has fallen on the individual's experience first.

Yet, the power of God—what the New Testament often calls the righteousness of God—has been poured into the world through Jesus Christ, making all things new. Individual lives certainly are transformed through this power; however, this is but part of God's saving grace which reclaims sovereignty over all the creation.

But we are still left with a problem. What is the relationship between the new creation promised, and the present creation given by God? And when is the new creation manifested? Only at the end of time, or beginning within history? Further, does this new creation come through the destruction of the old, or through its transformation?

Answers to such questions do not merely satisfy theological curiosity. Rather, our response is likely to influence our forms of spirituality and our witness to the social, political, and economic life of our world.

Forms of personal piety are likely to reveal understandings of these broader questions concerning creation and the new creation. And the possible options can best be understood by looking at the extremes.

A Flight from the World

Consider, as one example, Susan, an evangelical convert to Christian faith. Her past life, filled with sexual promiscuity, drug and alcohol abuse, spiteful jealousy of others, and bitterness toward her parents and well-intentioned friends, had brought her to utter despair and near self-destruction. But the message of God's love for her, and God's forgiveness of her sin through Christ, resulted in a conversion experience.

Susan sensed her life was filled with God's Spirit, making her a new person. This power was able to change old habits, feelings, and activities. All that she used to be was fading away; its remaining remnants simply evidenced the persistence of sin, or the power of the devil, fighting against her new life in Christ.

Spiritual life for her came to mean yielding to the external power of the Spirit that would possess her, helping her to live in this spiritual realm. Charismatic forms of worship nurtured her, emphasizing the surrender of her self to the Spirit. And her daily life became dominated by the desire to remain attentive and submitted to this spiritual presence.

She tried to keep the car radio tuned to the Christian station, feeling vaguely guilty for missing rock music. Her friendships focused around her Bible study group, and her only encounters with former friends consisted of strained attempts to witness to them about the new life she had discovered through Christ.

Her quiet fear was that she would not be strong enough to resist the temptations of her old life. So her defense became to cut off every possible tie to her past, regarding it all as hopelessly dominated by sin. Constantly, she reminded herself that this old self, and all associated with it, was dead.

For such a person, there is no continuity between the given creation and the new creation. The barrier between the two is total. Further, all of God's action comes through external, spiritual intervention rescuing one from the threatening powers found in the world. The new creation implies the eventual destruction of the old and depends, for now, on daily deliverance from its clutches.

Many Christians who totally reject such forms of evangelical personal piety still erect an analogous barrier between the present creation and the new creation when witnessing to the prophetic demands of God's justice. The inner personal, spiritual conflict pictured in the evangelical convert is projected corporately onto the church's relationship to the world.

The present structures and life in society are seen as hopelessly fallen, evil, and self-destructive. Often, a

radical social and economic analysis persuasively underscores such a judgment. One's main task is to be freed from
the clutches of this fallen creation. And the means may be
through some form of Christian community. The experience of such community serves as an expression of the new
creation.

Sectarian and monastic models emphasize withdrawal
from the social, political, economic and, at times, even geo-
graphical order of society. Attempts are made to sever con-
nections between the present order and the expressions of
the new creation. Life becomes organized and nurtured
from within such communities, and in resistance to the outside world.

In extreme examples, such communities view the
larger society as hopelessly marked for destruction. Their
interaction with it is confined to witnessing through acts of
dramatic resistance. Because the new creation has no continuity with the old, hopes are placed in political and even
quasi-spiritual intervention that will hasten the new creation as the present order crumbles and is destroyed.

The Other Extreme

Consider, however, the other extreme. Again, forms
of spirituality exemplify a basic assumption about this
creation and the new creation. In this case, picture a
thirty-five-year-old man, Gary, sincere in his Christian
commitment, but searching for forms of spirituality that
connect with his inward feelings and his intellectual sensi-
tivities. Raised in a strict Reformed background with a
conservative evangelical theology, Gary began searching
in college and through graduate school for a fresh understanding and experience of faith.

He rejected the rigid and protective theology of
his upbringing, learning in the university to search for
truth wherever it was to be found. Likewise, he spurned
the subcultural defensiveness of his church, believing instead that life in the larger society was not outside the

boundaries of God's grace. So he plunged in, began building a career, and enthusiastically embraced new social and cultural experiences.

An Episcopalian church in his gentrified neighborhood of renovated townhouses, popular with other younger professionals, became his new church home. The young rector's sermons were intellectually clever, with quotes and references from best-selling novels, tidbits from contemporary theologians, and subtle gibes at television evangelists. Moreover, the rector drove a BMW.

A study group in the church met Sunday evenings in homes to discuss Carl Jung over white wine and brie cheese. There, Gary began looking into his feelings, personality traits, and dreams as he sought to fashion a fresh spirituality. In discovering one's deepest self, and following its dictates, he came to believe, one was living in harmony with God.

The search, however, was not without pain and conflict with the external authority and spiritual insulation of his upbringing. Seeing a private therapist, however, seemed to help. During his sessions, he often remembered the quote, "all real life is meeting one's true self." Surely God must be present in such self-exploration. It became hard for him to distinguish between inner reflection and personal prayer.

The formality of the Episcopal liturgy suited him well, for it seemed to be a protection against the kind of external spiritual enthusiasm and shallow piety which he had come to distrust so deep within himself. Yet, he felt vaguely perplexed by the Eucharist itself; while he sensed it was critically important, he couldn't honestly explain to himself why this was so, or what he believed it actually meant.

For such a person, there is virtually no division between this creation and the new creation. The newness of life intended by God is all present within us, now. Searching, affirming, and celebrating our present life, with all its hidden potentials, enable this creation to be discovered as the new creation.

Removing this distinction has dramatic consequences for the gospel's relationship to the social, political, and economic structures of society. The most obvious result is a Constantinian marriage between the church and the public order. Founded on the conviction that God's common grace is present within the culture's structures and values, rather than in opposition to them, this stance seeks to affirm the goodness within the present order.

In its most extreme form, a given society is blindly embraced as embodying the new creation and undergirded by divine authority. History provides examples of both reactionary and utopian societies tempting the church with the claim that they constitute the new creation. Also, those Christian traditions weak in any eschatological vision—particularly some which grew over the centuries free from the dominant influence of Western culture—assume that since the creation has shared in God's redemption, the church's role is to liturgically celebrate this reality, rather than hope for unseen new heavens and a new earth.

The Nature of Christ

The relationship between this creation and the new creation eventually becomes a question about the nature of Jesus Christ. In fact, the early church's debates over Christology mirror, in certain ways, our contemporary confusion.

Some early church fathers—those associated with the opinions in Alexandria—fought to preserve the prominence of Christ's divinity. But in so doing, they risked severing his human identity. In a sense, they were reluctant to admit that the new creation in Christ could ever be united with the true humanity of this creation.

When adamant on this point, this position deteriorated into Docetism, asserting outright that Christ only had the appearance of a human, and denying his actual humanity. That fit nicely with Gnostic influences, which, to this day, infect forms of Christian piety, driving a wedge between external "spiritual" forces separated from and opposed to life in the world.

The early church leaders at Antioch began with the humanity of Jesus, stressing his historical existence and fighting to preserve the distinct reality of his human nature. In more extreme forms, they asserted that Jesus began merely as a human person and advanced, or was "adopted," as God's son, or else simply remained a creature. While the motivating concern of these theologians was their belief that the gulf between God and humanity couldn't be bridged, their eventual heritage seems to have been quite different.

Most modern Christology has sought to explore the nature of Christ "from below," rather than "from above." So it also begins with the concrete, historical humanity of Jesus of Nazareth. The effect has certainly been to underscore how the actual, human life of Jesus, in all its particularity, expressed the presence of God's Spirit. By implication, the spirituality of those who follow Jesus also becomes rooted below, in actual human experience, rather then imputed externally from above and overriding our humanity. Eventually, any transcendent dimensions seem eclipsed by the imminent realities. The new creation comes within this creation.

The early church's argument wasn't settled until the Council of Chalcedon in 451. One of its key points was that all the work of Christ came from his whole person, rather than from either his human or divine nature. Further, by then the church acknowledged that God couldn't fully redeem life if God had not fully embraced all of life.

The church today can well be guided by such wisdom. The promise of the new creation can no more be severed from this creation than Christ's divinity can be severed from his humanity. Whether in our personal lives or in the world, the new creation that comes takes its root within the realities of this life, not beyond them.

Yet, the new creation is truly new. It is never an affirmation of the present. Rather, this creation—its flesh and blood, its bread and wine—is invaded by the new creation, and opened to radical transformation.

Our spirituality, then, should never rest on external ecstasy offering escape from this life. But neither can it be equated with psychological self-discovery. Likewise, our witness to the world should not be limited to sectarian resistance. But neither can it ever be compromised by any form of blind Constantinianism.

Developing the theology of creation provides the means to guard the church against these various temptations and guide its spirituality and witness to embody the promise of the new creation. The biblical promise for the future is not for the present to be obliterated, and some wholly new reality to arise like a gnostic Phoenix out of the ashes. Rather, the promise is for God's original intended purposes of creation to be fulfilled.

God's embrace and redemption of all creation must never be confused with the sanctioning of given structures and the present order within society. Instead, the shape of all personal and corporate life in the world is opened to radical judgment and transformation through the "righteousness of God"—God's saving power active to restore and redeem all creation, through Christ, to its intended purpose of glorifying God.

In Jesus Christ, the creation is reclaimed as God's own. God's sovereignty over all its powers and principalities, and God's embrace of the created order as sustainer and Redeemer, are proclaimed through the cross and resurrection. The life of the Risen Christ, after all, invades our history; it enters into this creation; it encompasses the body; and it is present with grain from the fields, and wine from the vines.

8

Life and Death

Our exploration of the Bible's theology of creation and our responsibility toward the gift of the world's environment ultimately leads us to a consideration of life and death in the creation. And, here again, we must recover a foundation of biblical perspectives, often neglected in modern society and misunderstood in the church. It is essential that we do this in order to approach these ultimate questions with fresh vision and resourceful hope.

Modern culture, as the result of our unprecedented technological mastery over life, has radically challenged the biblical understanding of life and death and has obtained the power to bring about the death of the created world.

Even as technology threatens life on earth with death, technology also has been employed to conquer death. While effects of medical technology to further life and to prevent an untimely and tragic death are precious gifts for humanity, medical technology within most affluent countries is moving beyond this role. In the mindset of modern medicine, and in the intensive care units of our hospitals, death is no longer regarded as a natural process to which all humanity is subject. Rather, death is seen as an alien intruder.

Technology is enabling us to prolong the biological functions of life in ways that cause us to ask what is the meaning of life itself. In the words of one author who reflected on her experiences as a medical student, "A patient can commit no more grievous offense in a university hospital than to die. To die is spitefully and ungratefully to proclaim the inadequacy of doctors and their technology. In short, to die in a hospital is nothing short of heresy."

Modern humanity has achieved a technological mastery over creation unprecedented in the course of history. In the process we have usurped god-like powers, the powers of life and death themselves, in alienation from our dependence upon God. Our modern understanding of life and death has been secularized. Our culture attempts to understand these phenomena on their own terms, and without any reference to the Creator. This process of secularization, which the modern world has inherited from the Enlightenment and the scientific revolution, has changed all of our thinking and understanding.

In particular, life itself has been reduced to various phenomena or data which are subject to scrutiny and thereby explanation. Thus life becomes defined as basic biological functioning, or as chemical interactions, or as, in essence, the flow of genetic information. Undoubtedly such reductionism—and there are many varieties—has contributed to a moral callousness toward life.

If there is "one seamless web" of life that is God's gift of grace, then we must root our understanding and definition of such life in the God whom we know as Creator, Redeemer, and sustainer. Thus, we do well to turn to the resources of our faith and to the inspired wisdom of

Scripture in order to discover new, yet ancient, perspectives on the nature of life and death.

Life As God's Gift

When my son JonKrister was three-and-a-half, he asked me one day, "Why did God make people so that they die?" The modern world approaches the issue of life and death with this question of why we die. Our focus is on death, which seems like such an assault—so undeserved, so inexplicable, so offensive, and so threatening.

The Bible finds a different starting point. We find there little speculation about the origin of death. The Old Testament generally accepts this as simply a part of human existence. Rather, the Bible constantly marvels at life itself. Biblical reflections on these questions begin with this focus on life, rather than with anguished questions about death.

All life is regarded as God's gift, and rooted in God. For instance in Psalm 36:9 we read, "For with thee is the fountain of life. . . ." The Bible reflects no understanding of life apart from its relationship to God.

Life here means not merely human life. The same breath or Spirit (*ruach*) which moved over the earth at creation and breathed into humanity the breath of life is also sent forth as the Spirit which creates and sustains all things and renews the face of the ground (Ps. 104:30). This breath of life brings dust and spirit together into a single whole, a creation which exists only in relationship to its Creator. As Job 12:10 states, "In his hand is the life of every living thing and the breath of all mankind."

The modern conception of life differs markedly from these biblical insights. The culture's assumption is that our lives are our own. Individually, at least, life *is* at our own disposal. We possess within ourselves sovereignty over life, and each of us has the right to determine his or her life's destiny. Politics is frequently seen, at least within the Western democratic system, as enabling this pursuit of life and "happiness."

The charter documents of the United States speak directly of certain "inalienable rights," among these being life. An intrinsic right to existence is assumed. Even the language of the "right to life" movement is taken from these ideas.

Yet the Bible knows nothing of such a view. Life is never autonomous, independent, or sovereign unto itself. Most of all, it has no intrinsic right to be in and of itself. When considered carefully, in fact, the idea of "inalienable" and autonomous rights to life constitute a direct affront to the biblical declaration that God is the Creator, the author of life and death. The plain biblical fact is that life is not ours, but God's.

While marveling at the gift of life, the Old Testament regards death as part of a natural process. Since humanity was created from dust (Gen. 2:7) and would also return to dust (Gen. 3:19), humanity's mortality seems assumed by the Bible's creation accounts. Further, this appears to be the case without reference to the Fall. The warning in Genesis 3:3 was that if Adam and Eve ate of the fruit of the tree in the midst of the garden, then they would die. The meaning in Hebrew is clearly an immediate death. Yet, of course, they did not immediately die.

Throughout the Old Testament there is no general expectation of life continuing after death. Only a few faint glimmers of this hope can be found. This perspective casts fresh light on many Scriptures. The frequent references, for instance, to the Lord's salvation within the Old Testament refer not to the ongoing and eternal life of one's soul or spirit, but rather to a salvation which is known only within this life.

For instance, Psalm 37:39–40 reads:

> The salvation of the righteous is from the Lord;
>   he is their refuge in the time of trouble.
> The Lord helps them and delivers them;
>   he delivers them from the wicked, and saves them,
>   because they take refuge in him.

Salvation means, then, being kept safe, being delivered, being liberated, being restored to health, enjoying God's favor and blessing, and experiencing the continual endowment of life from God. Such salvation is the work of God. It comes because of God's steadfast love. It is the gift of grace, but it is the "saving" of life as known and experienced now.

While the New Testament adds considerably to this understanding of salvation, it is critical to keep in mind the Old Testament framework. In that perspective, God's work of salvation is not that which saves us after life, but rather, that which saves us with life, in this life.

Death presented a problem to those in the Old Testament, but only in certain circumstances. For instance, if one died in the prime of life, or if one died without having any children, death then seemed capricious and unjust. Isaiah pictures death as a threat because once dead, one can no longer praise God. "Death cannot praise thee; those who go down to the pit cannot hope for thy faithfulness. The living, the living, he thanks thee, as I do this day" (Isa. 38:18–19).

While the Old Testament presents the fact of living as a gift of God, it also declares that true life is found in a response to God's initiative and word. Life, then, means not simply living, but rather consists of a posture of attentiveness to God's presence and action. Thus we read in Deuteronomy 8:3 that humanity "does not live by bread alone, but . . . by everything that proceeds out of the mouth of the Lord."

The picture of life itself as responsiveness and obedience to God is strongly evident in passages such as Ezekiel 18:5–9. In stark terms this passage states that the person who obeys God's commands, specifically who does not "defile his neighbor's wife, . . . does not oppress any one, . . . gives his bread to the hungry and covers the naked with a garment, does not lend at interest or take any increase . . . he is righteous, he shall surely live, says the

Lord God." Similarly, the one who does commit such actions shall "surely die."

What is meant here by death and life? It may be comforting to believe that those who obey God live, and those who do not obey will die, but in fact life often presents us with the opposite reality, as it did also in the time of the Old Testament. While there seems to be no firm answer to that dilemma within the Old Testament, its words consistently suggest that life itself is more than just living; life consists of right relationships.

The alternative is an estrangement which is rooted in our own autonomy. We desire to live unto ourselves, believing that our lives are our own intrinsic possession and natural right. This stance makes death become a terrible threat. Death ends this autonomy, and death finally defeats human attempts to claim the ownership of one's life. Thus, instead of death being a natural end to a life that always depends solely on God, death becomes the chief threat to our identity and denies the meaning in our existence.

Amos said, "Seek the Lord and live" (Amos 5:6). Such seeking of God—responding to the gift of life and acknowledging this relationship—is, in fact, life itself. Living in any other way denies life.

Jesus Christ, the Life of the World

The Old Testament perspectives concerning life and death are brought to fullness in the New Testament, particularly in the understanding of Jesus as both God's Messiah and Lord, meaning the one who reigns over all life. As stated previously, the prologue to the Gospel of John reads, "All things were made through him, and without him was not anything made that was made. In him was life, and the life was the light of all." This is one of several New Testament declarations that all creation and life come into being through Christ and are held together by Christ.

John describes Christ in symbols linked with the essence of life. Thus, Christ is the living water (John 4:10) and the "bread of life" (6:35, 48). Further, this is the bread of God "which comes down from heaven, and gives life to the world" (6:33). And later, "the bread which I shall give for the life of the world is my flesh" (6:51).

As in the Old Testament, the understanding of life is neither spiritualized nor individualized. It is expressed as coming not simply from Christ's spirit, but rather from Christ's flesh, from this same union of dust and breath. Life is not individualized; rather, Christ gives life to the whole world. All that exists in creation is nourished by the sacrificial flesh and blood of the Lord of life.

In the Gospel of John this life is also described as "eternal life." However, various descriptions of life and the phrase "eternal life" are used almost interchangeably. Eternal life is simply another way of describing the life that has come in Jesus Christ, and that comes to us as well as to all the whole creation. It is not, therefore, a part of life known only at the end of time, but rather a quality of the life which was exhibited by Christ and which is made known to us now.

While the Johannine writings provide a profound understanding of life, Paul presents us with a detailed, in-depth perspective of the meaning of death. In Romans 5:12 Paul states that death enters the world through sin. Yet in light of the Old Testament, more is meant here than merely the entrance of biological mortality into the world. I have already argued that such biological mortality seems present from the beginning of creation.

Paul constantly sees sin and death as one and the same. Death simply becomes the living out of this stance of disobedience, rather than mere biological mortality. Thus, death becomes understood as the power of rebellion against the purposes of God's life for the world.

This power has apparent dominion. The powers and principalities frequently referred to in the New Testament are embodied in all those forces which attempt to structure life apart from the sovereignty of God. Thus, the early church spoke of the dominion of sin, death, and the devil, meaning all those powers that would govern life in rebellion to God.

Such power attempts to gain hold over all creation. Humanity's attempt to rule over life and creation and to be "like God" reflects this state of rebellion and alienation. Thus, such death is a state of living. It is a power within the creation, and it is inseparable from sin.

When Paul says, therefore, that "death entered the world through sin," the theological meaning is that rebellion against God's sovereignty results in the condition of living in separation from God. Such a condition is described as death. Biological death in such a condition threatens to make that separation permanent.

Romans goes on to declare that nothing separates us from the love of God (Rom. 8:35), including the power of death. In the classic understanding of redemption, it is precisely this power of sin, death, and the devil which is overcome by the cross and resurrection. Christ's work of redemption restores life to the whole creation. In the words of Luther, "Curse yields to blessing."

The world is reclaimed by God through Christ, and knows true life. Such life is the life that overcomes all those powers that would separate creation from God's sovereignty. Even though this victory is complete only at the end of time, yet its triumph is certain. The reign of the powers of sin, death, and the devil has been broken.

Given this new reality, physical death no longer need be an enemy. It has been "swallowed up in victory" (1 Cor. 15:54). No longer is physical death a threat to retaining our autonomy. No longer is it a completion of our separation from God. No longer does it signify a reign of alienation and rebellion. Rather, it becomes simply an entry into the full presence of God, and a natural occurrence which holds no power to separate us from God's love.

The idea of a general resurrection at the end of time was already present within the Jewish community at the time of Jesus. The Sadducees, who did not believe in the resurrection, questioned Jesus on that basis.

What was unique in the resurrection of Jesus, however, was not merely a rising or recovery from life only to die again, as in the case of Lazarus. Rather, the resurrection of Christ signified that the new age of God's reign and the power of God's life—this final act of God to redeem the creation—begins now in the midst of history.

We do not wait for the end of history, or the end of the world, in order for the power of this resurrection life which defeats the powers of death to be present. Rather, Christ's resurrection is a demonstration that this triumphant power has already entered into the world.

Furthermore, it has entered into us. This life of the world to come is a life in which we now participate. To live in Christ and to live through him indicate that our lives are opened to the triumph of Christ over the powers of sin and death. In the words of Roman 6:13 we are brought "from death to life," and we reign in life through Jesus Christ. All this is through God's grace. Life, which is initially known as God's gift, comes in its fullness through the action of God in Jesus Christ.

Last Easter my daughter Karis was ill. That morning my son JonKrister asked, "Has Karis been resurrected from her ear infection?" Theologically, he had a point. The promise of resurrection is linked to the hope of healing and restoration for all the creation. The destruction and alienation that come through death—separation from God—has finally been overcome. The resurrection of Christ means that in this life we taste the first fruits of that full life intended by God for the world. It was this truth that enabled early Christians to face confidently their own death in martyrdom. Their suffering and sacrifice were based on the assurance that this new life has already triumphed.

Finally, for the church of the New Testament period, two concrete experiences seem linked to the reality of this resurrection life in their midst—healing and community. Healings in Acts are regarded as manifestations of this resurrection power and of the establishment of God's reign in the world. That is why such healings were so threatening to those holding established political and religious power.

Resurrection life also meant community. The taste of this new life was known not in privatized spiritual ecstasy, but in fellowship with one another. Here is where this new life was, and is, known.

We all would do well to reconsider our understanding today of life and death in light of biblical perspectives. If we have already passed from death to life, then we are to partake in the fresh power and spirit shaping the community of God's people, and offering life to the world.

The life which we offer is the resurrected Christ, who is the life of the world. Since the first Easter, Christians have stressed the resurrection of Christ's body. The post-resurrection accounts of the risen Lord with the disciples make a point of emphasizing that his body was real. He ate with them, and he asked Thomas to feel the wounds in his hands and side.

The resurrection of the body emphasizes that the material, the flesh and blood of this world, is united with the Spirit and transformed into the life of the world to come. For a theology of creation and the stance of Christians toward the created world, this truth is critical. In a time of ecological emergency, the church can offer to the world a hope that is rooted in the power of God to bring new life into all that has been created. And that power has already entered into history; it is amongst us now, working to redeem and reconcile all that is broken so the new creation can be realized.

Epilogue

Fly-fishing As a
Spiritual Discipline

I grew up praying in the suburban evangelical subculture of the 1950s. Prayer was to be regular, within a daily quiet time, as well as spontaneous, for help and guidance throughout the day. Its most frequent purpose was to ask for something that I wanted, or to keep from doing something I didn't want to do. Like most of life in early adolescence, prayer was narcissistic.

I grew up fishing in Williams Bay, Wisconsin, on Lake Geneva. Fishing was to be regular, at dawn or dusk, with sporadic forays in the day if life was boring. Its chief purpose was to catch what I wanted to eat, usually perch, small bass, and sunfish, and not to catch what I didn't want, particularly catfish.

Uncle Bob—actually my grandfather's brother—taught me how to fish. You took the boat straight out from the house until, when you looked back, the flagpost was in front of the chimney in your line of sight. Then you were right at the "drop off" where schools of fish liked to be. The line and hook, with night crawlers gathered the previous evening, went down until the sinker hit the bottom, and then was lifted a foot; having done that, you waited.

We would catch a dozen or so "keeping size" on a good day, carry them to a table behind the well house, hit their heads with the butt of a knife, scale them, chop off their heads and tails, clean out their guts while we kept the bees away, and then take them into the kitchen to be fried up in butter with dinner. The point of fishing was to catch and eat fish.

Several people taught me to pray—my parents, of course, but also church youth leaders, my pastor, my Young Life Club leader, and a stream of devotional books about prayer. Prayer requests were dug up, both for me and others. And the prayers were sent out, or up, until they landed. Then I waited. The point of praying was to get answers and results.

Eventually the well of spiritual enthusiasm ran dry. Prayer no longer produced the same sort of inner excitement. And the answers seemed more elusive. The old formulas didn't function. A daily "quiet time" became a relic of past pietism. Faith and commitment remained strong. But I read *Are You Running with Me Jesus?* and prayers by Michael Quoist while sensing that my earlier "personal walk with the Lord" had come to an inner dead end.

This was about the same time that I lost interest in fishing. Uncle Bob had moved to Florida. At the Williams Bay house, fishing poles, lines, and tackle were harder to find, and always tangled anyway. Digging worms seemed a chore. Buying them was a betrayal of the tradition. And I was certain the fish were scarcer and smaller than I had remembered. It was all too much bother for too little result. The excitement had gone out of my fishing as well as my prayer.

The rolling hills by Berryville, Virginia, beckoned me for a retreat away from the daily, crusading pace of Washington D.C. in the 1970s. My faith had me fighting to end a war, and working for justice, seventy to eighty hours a week. Changing the world was leaving little time for prayer or fishing. But the Church of the Saviour had reopened my "inward journey." So I went to the guest house of the Trappist monastery by Berryville.

Like many Protestants alienated from praying for parking places downtown, but sensing an inner hunger, I had discovered Thomas Merton. And at Berryville, another Trappist, Father Stephen, taught me how to pray. Results weren't the point. Presence was—the presence of the whole self to God. And when open to God, we are opened to the world—its suffering, pain, and hope.

Getting wasn't the point either. Prayer was freed from production. No longer did I think up a prayer list. Rather, prayer meant erasing it. And listening. Emptying the self of its enduring narcissism and discovering the presence of Christ within me, others, and in the world.

Discipline was essential with regular times, and even techniques. Journaling, centering, and contemplation formed a new vocabulary of piety. I listened to my teacher, and watched him "practice the presence." Then I lived with the monks for six weeks to absorb their craft, which was more like art. The point of prayer became prayer.

Vocation, which I now named "calling," shifted. And my inward journey led me not only to God, but to marriage. After that came community—intense community, with ministry to and life with the poor. And the crusading continued, now against nuclear war. I began forgetting the days at Berryville. I led retreats, but never went on any. When I finally did, I was called to journey on, and in. And out of Washington, D.C.

The mountains of Montana now beckoned me. And there, besides another community which welcomed me, was a graduate of Hope College and Western Seminary who spent his weekends, and parts of his week, fly-fishing. He offered to teach me, and became my novice-master. I listened, read, and went with him to watch, and then practice.

He knew intimately the life of insects on the water, when they would hatch, and how they looked. Wading into the river, he was present to it, and nothing more. The gracefully looping flyline, tipped by a fly which he had tied, placed it onto the water as though the fly were part of its life. He watched, and listened, and emptied himself of all else.

I loved to eat trout. But for him, none was "keeping size." He was a catch-and-release fly-fisherman. He once confessed to keeping a few when his in-laws had come. But such sins no longer tempted him. For him, fly-fishing was freed from producing a catch. He went to be there. The point of fishing was fishing. And his craft was more of an art.

Then children came. I don't know yet whether they have done more damage to my prayer or my fly-fishing. Neither has fared well. Time, even in Montana, is pressed in once again by the calls not only of community and crusades, but also of kids.

The first day of fishing season this year, a friend from the church community called, suggesting that we go out on the Bitterroot River. When he prays, I think he expects answers. And when he fishes, he expects fish. And I hadn't eaten fresh trout in a long time. He suggested ten to six, and I, noon to four. As the new father of a two-month-old baby girl, he quickly agreed.

During the first fifteen minutes he caught a nice-size rainbow trout. Thinking of fish on the grill that night, I cast earnestly over rises forty feet away. A trout rose to my fly; I jerked the line impulsively, and it slipped off. After more frustration over similar results, I went to change flies. And then I noticed that the hook on the fly I was using was barbless. It had been a gift from my novice-master, anticipating a time when I would want to try fishing with hooks designed not to keep fish.

Now even more determined and discontent, I saw my friend catch his second rainbow. I headed over to where he had caught the first one, tied on a fly that matched the hatch, and began almost fitful, aggressive casts reaching for good water. The wind was blowing strong; my technique was being forced. As a back cast started, I saw the line come toward my face.

The small hook on the fly I had selected worked perfectly. It was imbedded firmly into the inside of my lower lip. And it was barbed.

The doctor at the emergency clinic marveled at the catch. "Let's have this one stuffed and mounted." I lay there pondering a contemplative question once suggested by Henri Nouwen, "What is the meaning of what seems to be?"

The insurance company sent back the claim form for the clinic bill with a note saying , "Please include a detailed description of your accident including how it occurred." I should have responded, "I forgot how to pray."

Appendix

A vast literature on all phases of ecology has been developing in recent years. While a profound appreciation of created reality has traditionally characterized Christian theology, its allegedly negative and arrogant attitude toward nature has come under sharp attack by contemporary critics like Lynn White, Jr. Motivated not only to answer unjustified structures but also to explore more thoroughly what might be called a biblical ecology, Christians have been adding significantly to the literature on this subject.

The essays and excerpts here included are a mere cross-section of the voluminous material now available. A sort of postscript to Wesley Granberg-Michaelson's text, their purpose is to give additional insight and clarified understanding to concerned readers.

The inclusion of a particular viewpoint does not, however, imply endorsement. It simply means that the argument or perspective being set forth is an important statement which merits attention even though it may elicit disagreement. For instance, White's article, often reprinted since it was published over twenty years ago, is a landmark indictment against Western Christianity for, so White charges, inculcating within its adherents a rapacious irresponsibility toward our planet and its resources.

In the third quarter of our century, Francis Schaeffer from L'Abri, his spiritual retreat and study center in Huemoz, Switzerland, exerted a powerful influence on evangelical thought and action. His book, *Pollution and the Death of Man*, was a telling rejoinder to White's thesis.

In his inaugural address as professor of Old Testament at Wesley Theological Seminary in Washington, D.C., Bruce C. Birch explored the threefold relationship among "Nature, Humanity and Biblical Theology." In effect, he sketched a relational theology of nature.

Approaching this issue from another angle of vision, Vincent Rossi—director general of the Holy Order of MANS, a Christian brotherhood, and a perceptive writer on ecological matters—argues the case for "Theocentrism: The Cornerstone of Christian Ecology."

James A. Rimbach, professor of theology in Concordia Seminary, Kowloon, Hong Kong, draws out the planetary and cosmic implications of the apostle Paul's teaching in chapter 8 of his Letter to the Romans.

In conclusion, H. Paul Santmire of Wellesley College, Wellesley, Massachusetts, who has been in the vanguard of faith-oriented ecological thinking, highlights "God's Joyous Valuing of Nature."

Obviously, these are only a sampling. But they may, together with the text of this volume in the series, Issues of Christian Conscience, stimulate reflection and research, concern and action with respect to the formidable problem of ethical earthkeeping.

THE EDITOR

The Historical Roots of Our Ecologic Crisis

Lynn White, Jr.

A conversation with Aldous Huxley not infrequently put one at the receiving end of an unforgettable monologue. About a year before his lamented death he was discoursing on a favorite topic: man's unnatural treatment of nature and its sad results. To illustrate his point he told how, during the previous summer, he had returned to a little valley in England where he had spent many happy months as a child. Once it had been composed of delightful grassy glades; now it was becoming overgrown with unsightly brush because the rabbits that formerly kept such growth under control had largely succumbed to a disease, myxamatosis, that was deliberately introduced by the local farmers to reduce the rabbits' destruction of crops. Being something of a Philistine, I could be silent no longer, even in the interests of great rhetoric. I interrupted to point out that the rabbit itself had been brought as a domestic animal to England in 1176, presumably to improve the protein diet of the peasantry.

All forms of life modify their contexts. The most spectacular and benign instance is doubtless the coral polyp. By serving its own ends, it had created a vast undersea world favorable to thousands of other kinds of animals and plants. Ever since man became a numerous species he has affected his environment notably. The hypothesis that his fire-drive method of hunting created the world's great grasslands and helped to exterminate the monster mammals of the Pleistocene from much of the globe is plausible, if not proved. For six millennia at least, the banks of the lower Nile have been a human artifact rather than the swampy African jungle which nature, apart from man, would have made it. The Aswan Dam, flooding five thousand square miles, is only the latest stage in a long process. In many regions terracing or irrigation, overgrazing, the cutting of forests by

Romans to build ships to fight Carthaginians or by Crusaders to solve the logistics problems of their expeditions, have profoundly changed some ecologies. Observation that the French landscape falls into two basic types, the open fields of the north and the *bocage* of the south and west, inspired Marc Bloch to undertake his classic study of medieval agricultural methods. Quite unintentionally, changes in human ways often affect nonhuman nature. It has been noted, for example, that the advent of the automobile eliminated huge flocks of sparrows that once fed on the horse manure littering every street.

The history of ecologic change is still so rudimentary that we know little about what really happened, or what the results were. The extinction of the European aurochs as late as 1627 would seem to have been a simple case of overenthusiastic hunting. On more intricate matters it often is impossible to find solid information. For a thousand years or more the Frisians and Hollanders have been pushing back the North Sea, and the process is culminating in our own time in the reclamation of the Zuider Zee. What, if any, species of animals, birds, fish, shore life, or plants have died out in the process? In their epic combat with Neptune have the Netherlanders overlooked ecological values in such a way that the quality of human life in the Netherlands has suffered? I cannot discover that the questions have ever been asked, much less answered.

People, then, have often been a dynamic element in their own environment, but in the present state of historical scholarship we usually do not know exactly when, where, or with what effects man-induced changes came. As we enter the last third of the twentieth century, however, concern for the problem of ecologic backlash is mounting feverishly.

Natural science, conceived as the effort to understand the nature of things, had flourished in several eras among several peoples. Similarly there had been an age-old accumulation of technological skills, sometimes growing rapidly, sometimes slowly. But it was not until about four generations ago that Western Europe and North America arranged a marriage between science and technology, a union of the theoretical and the empirical approaches to our natural environment. The emergence in widespread practice of the Baconian creed that scientific knowledge means technological power over nature can scarcely be dated before about 1850, save in the chemical industries, where it is anticipated in the eighteenth century. Its acceptance as a normal pattern of action may mark the greatest event in human history since the invention of agriculture, and perhaps in nonhuman terrestrial history as well.

Almost at once the new situation forced the crystallization of the novel concept of ecology; indeed, the word *ecology* first appeared in the English language in 1873.

Today, less than a century later, the impact of our race upon the environment has so increased in force that it has changed in essence. When the first cannons were fired, in the early fourteenth century, they affected ecology by sending workers scrambling to the forests and mountains for more potash, sulfur, iron ore, and charcoal, with some resulting erosion and deforestation. Hydrogen bombs are of a different order: a war fought with them might alter the genetics of all life on this planet.

By 1285 London had a smog problem arising from the burning of soft coal, but our present combustion of fossil fuels threatens to change the chemistry of the globe's atmosphere as a whole, with consequences which we are only beginning to guess. With the population explosion, the carcinoma of planless urbanism, the now geological deposits of sewage and garbage; surely no creature other than man has ever managed to foul its nest in such a short order.

There are many calls to action; but specific proposals, however worthy as individual items, seem too partial, palliative, negative: ban the bomb, tear down the billboards, give the Hindus contraceptives and tell them to eat their sacred cows. The simplest solution to any suspect change is, of course, to stop it, or, better yet, to revert to a romanticized past: make those ugly gasoline stations look like Anne Hathaway's cottage or (in the Far West) like ghost-town saloons. The "wilderness area" mentality invariably advocates deep-freezing an ecology, whether San Gimignano or the High Sierra, as it was before the first Kleenex was dropped. But neither atavism nor prettification will cope with the ecologic crisis of our time.

What shall we do? No one yet knows. Unless we think about fundamentals, our specific measures may produce new backlashes more serious than those they are designed to remedy.

As a beginning we should try to clarify our thinking by looking, in some historical depth, at the presuppositions that underlie modern technology and science. Science was traditionally aristocratic, speculative, intellectual in intent; technology was lower-class, empirical, action-oriented. The quite sudden fusion of these two, towards the middle of the nineteenth century, is surely related to the slightly prior and contemporary democratic revolutions which, by reducing social barriers, tended to assert a functional unity of brain and hand. Our ecologic crisis is the product of an emerging, entirely novel, democratic culture. The issue is whether a democratized world can survive its own implications. Presumably we cannot unless we rethink our axioms.

The Western Traditions of Technology and Science

One thing is so certain that it seems stupid to verbalize it: both modern technology and modern science are distinctly *Occidental*. Our technology has absorbed elements from all over the world, notably from China; yet everywhere today, whether in Japan or in Nigeria, successful technology is Western. Our science is the heir to all sciences of the past, especially perhaps to the work of the great Islamic scientists of the Middle Ages, who so often outdid the ancient Greeks in skill and perspicacity: al-Razi in medicine, for example; or Ibn al-Haytham in optics; or Omar Khayyam in mathematics. Indeed, not a few works of such geniuses seem to have vanished in the original Arabic and to survive only in medieval Latin translations that helped to lay the foundations for later Western developments. Today, around the globe, all significant science is Western in style and method, whatever the pigmentation or language of the scientists.

A second pair of facts is less well recognized because they result from quite recent historical scholarship.

The leadership of the West, both in technology and in science, is far older than the so-called Scientific Revolution of the seventeenth century or the so-called Industrial Revolution of the eighteenth century. These terms are in fact outmoded and obscure the true nature of what they try to describe— significant stages in two long and separate developments.

By A.D. 1000 at the latest—and perhaps, feebly, as much as two hundred years earlier—the West began to apply water power to industrial processes other than milling grain. This was followed in the late twelfth century by the harnessing of wind power. From simple beginnings, but with remarkable consistency of style, the West rapidly expanded its skills in the development of power machinery, labor-saving devices, and automation. Those who doubt should contemplate that most monumental achievement in the history of automation: the weight-driven mechanical clock, which appeared in two forms in the early fourteenth century.

Not in craftsmanship but in basic technological capacity, the Latin West of the later Middle Ages far outstripped its elaborate, sophisticated, and esthetically magnificent sister cultures, Byzantium and Islam. In 1444 a great Greek ecclesiastic, Bessarion, who had gone to Italy, wrote a letter to a prince in Greece. He is amazed by the superiority of Western ships, arms, textiles, glass. But above all he is astonished by the spectacle of waterwheels sawing timbers and pumping the bellows of blast furnaces. Clearly, he had seen nothing of the sort in the Near East.

By the end of the fifteenth century the technological superiority of Europe was such that its small, mutually hostile nations could spill out over all the rest of the world, conquering, looting, and colonizing. The symbol of this technological superiority is the fact that Portugal, one of the weakest states of the Occident, was able to become, and to remain for a century, mistress of the East Indies. And we must remember that the technology of Vasco da Gama and Albuquerque was built by pure empiricism, drawing remarkably little support or inspiration from science.

In the present-day vernacular understanding, modern science is supposed to have begun in 1543, when both Copernicus and Vesalius published their great works. It is no derogation of their accomplishments, however, to point out that such structures as the *Fabrica* and the *De revolutionibus* do not appear overnight. The distinctive Western tradition of science, in fact, began in the late eleventh century with a massive movement of translation of Arabic and Greek scientific works into Latin. A few notable books—Theophrastus, for example—escaped the West's avid new appetite for science, but within less than two hundred years effectively the entire corpus of Greek and Muslim science was available in Latin, and was being eagerly read and criticized in the new European universities. Out of criticism arose new observation, speculation, and increasing distrust of ancient authorities.

By the late thirteenth century Europe had seized global scientific leadership from the faltering hands of Islam. It would be as absurd to deny the profound originality of Newton, Galileo, or Copernicus as to deny that of the fourteenth-century scholastic scientists like Buridan or Oresme on whose work they built. Before the eleventh century, science scarcely existed in the Latin West, even in Roman times. From the eleventh century onward, the scientific sector of Occidental culture has increased in a steady crescendo.

Since both our technological and our scientific movements got their start, acquired their character, and achieved world dominance in the Middle Ages, it would seem that we cannot understand their nature or their present impact upon ecology without examining fundamental medieval assumptions and developments.

Medieval View of Man and Nature

Until recently, agriculture has been the chief occupation even in "advanced" societies; hence, any change in methods of tillage has much importance. Early plows, drawn by two oxen, did not normally turn the sod but merely scratched it. Thus, cross-plowing was needed and fields tended to be squarish. In the fairly light soils and semiarid climates of the Near East and Mediterranean, this worked well. But such a plow was inappropriate to the wet climate and often sticky soils of northern Europe.

By the latter part of the seventh century after Christ, however, following obscure beginnings, certain northern peasants were using an entirely new kind of plow, equipped with a vertical knife to cut the line of the furrow, a horizontal share to slice under the sod, and a moldboard to turn it over. The friction of this plow with the soil was so great that it normally required not two but eight oxen. It attacked the land with such violence that cross-plowing was not needed, and fields tended to be shaped in long strips.

In the days of the scratch-plow, fields were distributed generally in units capable of supporting a single family. Subsistence farming was the presupposition. But no peasant owned eight oxen: to use the new and more efficient plow, peasants pooled their oxen to form large plow-teams, originally receiving (it would appear) plowed strips in proportion to their contribution. Thus, distribution of land was based no longer on the needs of a family but, rather, on the capacity of a power machine to till the earth.

Man's relation to the soil was profoundly changed. Formerly man had been part of nature; now he was the exploiter of nature. Nowhere else in the world did farmers develop any analogous agricultural implement. Is it coincidence that modern technology, with its ruthlessness toward nature, has so largely been produced by descendants of these peasants of northern Europe?

This same exploitive attitude appears slightly before A.D. 830 in western illustrated calendars. In older calendars the months were shown as passive personifications. The new Frankish calendars, which set the style for the Middle Ages, are very different: they show men coercing the world around them—plowing, harvesting, chopping trees, butchering pigs. Man and nature are two things, and man is master.

These novelties seem to be in harmony with larger intellectual patterns. What people do about their ecology depends on what they think about themselves in relation to things around them. Human ecology is deeply conditioned by beliefs about our nature and destiny—that is, by religion. To western eyes this is very evident in, say, India or Ceylon. It is equally true of ourselves and of our medieval ancestors.

The victory of Christianity over paganism was the greatest psychic revolution in the history of our culture. It has become fashionable today to say that, for better or worse, we live in "the post-Christian age." Certainly the forms of our thinking and language have largely ceased to be Christian, but to my eye the substance often remains amazingly akin to that of the past. Our daily habits of action, for example, are dominated by an implicit faith in perpetual progress which was unknown either to Greco-Roman antiquity or to the Orient. It is rooted in, and is indefensible apart from, Judeo-Christian teleology. The fact that Communists share it merely helps to show what can be demonstrated on many other grounds: that Marxism, like Islam, is a Judeo-Christian heresy. We continue today to live, as we have lived for about seventeen hundred years, very largely in a context of Christian axioms.

What did Christianity tell people about their relations with the environment?

While many of the world's mythologies provide stories of creation, Greco-Roman mythology was singularly incoherent in this respect. Like Aristotle, the intellectuals of the ancient West denied that the visible world had had a beginning. Indeed, the idea of a beginning was impossible in the framework of their cyclical notion of time.

In sharp contrast, Christianity inherited from Judaism not only a concept of time as nonrepetitive and linear but also a striking story of creation. By gradual stages a loving and all-powerful God had created light and darkness, the heavenly bodies, the earth and all its plants, animals, birds, and fishes. Finally, God had created Adam and, as an afterthought, Eve to keep man from being lonely. Man named all the animals, thus establishing his dominance over them. God planned all of this explicitly for man's benefit and rule: no item in the physical creation had any purpose save to serve man's purposes. And, although man's body is made of clay, he is not simply part of nature: he is made in God's image.

Especially in its western form, Christianity is the most anthropocentric religion the world has seen. As early as the second century both Tertullian and Saint Irenaeus of Lyons were insisting that when God shaped Adam he was foreshadowing the image of the incarnate Christ, the Second Adam. Man shares, in great measure, God's transcendence of nature. Christianity, in absolute contrast to ancient paganism and Asia's religions (except, perhaps, Zoroastrianism), not only established a dualism of man and nature but also insisted that it is God's will that man exploit nature for his proper ends.

At the level of the common people this worked out in an interesting way. In antiquity every tree, every spring, every stream, every hill had its own *genius loci*, its guardian spirit. These spirits were accessible to men, but were very unlike men; centaurs, fauns, and mermaids show their ambivalence. Before one cut a tree, mined a mountain, or dammed a brook, it was important to placate the spirit in charge of that particular institution, and to keep it placated. By destroying pagan animism, Christianity made it possible to exploit nature in a mood of indifference to the feelings of natural objects.

It is often said that for animism the church substituted the cult of saints. True; but the cult of saints is functionally quite different from animism. The saint is not *in* natural objects; he may have special shrines, but his citizenship is in heaven. Moreover, a saint is entirely a man; he can be approached in human terms.

In addition to saints, Christianity of course also had angels and demons inherited from Judaism and perhaps, at one remove, from Zoroastrianism. But these were all as mobile as the saints themselves. The spirits *in* natural objects, which formerly had protected nature from man, evaporated. Man's effective monopoly on spirit in this world was confirmed, and the old inhibitions to the exploitation of nature crumbled.

When one speaks in such sweeping terms, a note of caution is in order. Christianity is a complex faith, and its consequences differ in differing contexts. What I have said may well apply to the medieval West, where in fact technology made spectacular advances. But the Greek East, a highly civilized realm of equal Christian devotion, seems to have produced no marked technological innovation after the late seventh century, when Greek fire was invented.

The key to the contrast may perhaps be found in a difference in the tonality of piety and thought which students of comparative theology find between the Greek and the Latin Churches. The Greeks believed that sin was intellectual blindness, and that salvation was found in illumination, orthodoxy— that is, clear thinking. The Latins, on the other hand, felt that sin was moral evil, and that salvation was to be found in right conduct. Eastern theology has been intellectualist. Western theology has been voluntarist. The Greek saint contemplates; the Western saints acts. The implications of Christianity for the conquest of nature would emerge more easily in the Western atmosphere.

The Christian dogma of creation, which is found in the first clause of all the Creeds, has another meaning for our comprehension of today's ecologic crisis.

By revelation, God had given man the Bible, the Book of Scripture. But since God had made nature, nature also must reveal the divine mentality. The religious study of nature for the better understanding of God was known as natural theology. In the early church, and always in the Greek East, nature was conceived primarily as a symbolic system through which God speaks to men: the ant is a sermon to sluggards; rising flames are the symbol of the soul's aspiration. This view of nature was essentially artistic rather than scientific. While Byzantium preserved and copied great numbers of ancient Greek scientific texts, science as we conceive it could scarcely flourish in such an ambience.

However, in the Latin West by the early thirteenth century natural theology was following a very different bent. It was ceasing to be the decoding of the physical symbols of God's communication with man and was becoming the effort to understand God's mind by discovering how his creation operates. The rainbow was no longer simply a symbol of hope first sent to Noah after the Deluge: Robert Grosseteste, Friar Roger Bacon, and Theodoric of Freiberg produced startlingly sophisticated work on the optics of the rainbow, but they did it as a venture in religious understanding.

From the thirteenth century onward, up to and including Leibnitz and Newton, every major scientist, in effect, explained his motivations in religious terms. Indeed, if Galileo had not been so expert an amateur theologian he would have got into far less trouble: the professionals resented his intrusion. And Newton seems to have regarded himself more as a theologian than as a scientist. It was not until the late eighteenth century that the hypothesis of God became unnecessary to many scientists.

It is often hard for the historian to judge, when men explain why they are doing what they want to do, whether they are offering real reasons or merely culturally acceptable reasons. The consistency with which scientists during the long formative centuries of western science said that the task and the reward of the scientist was "to think God's thoughts after him" leads one to believe that this was their real motivation. If so, then modern western science was cast in a matrix of Christian theology. The dynamism of religious devotion, shaped by the Judeo-Christian dogma of creation, gave it impetus.

An Alternative Christian View

We would seem to be headed toward conclusions unpalatable to many Christians. Since both *science* and *technology* are blessed words in our contemporary vocabulary, some may be happy at the notions, first, that, viewed historically, modern science is an extrapolation of natural theology and, second, that modern technology is at least partly to be explained as an Occidental, voluntarist realization of the Christian dogma of man's transcendence of, and rightful mastery over, nature. But, as we now recognize, somewhat over a century ago science and technology—hitherto quite separate activities—joined to give mankind powers which, to judge by many of the ecologic effects, are out of control. If so, Christianity bears a huge burden of guilt.

I personally doubt that disastrous ecologic backlash can be avoided simply by applying to our problems more science and more technology. Our science and technology have grown out of Christian attitudes toward man's relation to nature which are almost universally held not only by Christians and neo-Christians but also by those who fondly regard themselves as post-Christians. Despite Copernicus, all the cosmos rotates around our little globe. Despite Darwin, we are *not*, in our hearts, part of the natural process. We are superior to nature, contemptuous of it, willing to use it for our slightest whim.

The newly elected Governor of California, like myself a churchman but less troubled than I, spoke for the Christian tradition when he said (as is alleged), "when you've seen one redwood tree, you've seen them all." To a Christian a tree can be no more than a physical fact. The whole concept of the sacred grove is alien to Christianity and to the ethos of the West. For nearly two millennia Christian missionaries have been chopping down sacred groves, which are idolatrous because they assume spirit in nature.

What we do about ecology depends on our ideas of the man-nature relationship. More science and more technology are not going to get us out of the present ecologic crisis until we find a new religion, or re-think our old one. The beatniks, who are the basic revolutionaries of our time, show a sound instinct in their affinity for Zen Buddhism, which conceives of the man-nature relationship as very nearly the mirror image of the Christian view. Zen, however, is as deeply conditioned by Asian history as Christianity is by the experience of the West, and I am dubious of its viability among us.

Possibly we should ponder the greatest radical in Christian history since Christ: Saint Francis of Assisi. The prime miracle of Saint Francis is the fact that he did not end at the stake, as many of his left-wing followers did. He was so clearly heretical that a general of the Franciscan Order, Saint Bonaventure, a great and perceptive Christian, tried to suppress the early accounts of Franciscanism. The key to an understanding of Francis is his belief in the virtue of humility—not merely for the individual but for man as a species. Francis tried to depose man from his monarchy over creation and set up a democracy of all God's creatures. With him the ant is no longer simply a homily for the lazy, flame a sign of the thrust of the soul toward union with God; now they are Brother Ant and Sister Fire, praising the Creator in their own ways as Brother Man does in his.

Later commentators have said that Francis preached to the birds as a rebuke to men who would not listen. The records do not read so: he urged the little birds to praise God, and in spiritual ecstasy they flapped their wings and chirped rejoicing. Legends of saints, especially the Irish saints, had long told of their dealings with animals but always, I believe, to show their human dominance over creatures. With Francis it is different. The land around Gubbio in the Apennines was being ravaged by a fierce wolf. Saint Francis, says the legend, talked to the wolf and persuaded him of the error of his ways. The wolf repented, died in the odor of sanctity, and was buried in consecrated ground.

What Sir Steven Ruciman calls "the Franciscan doctrine of the animal soul" was quickly stamped out. Quite possibly it was in part inspired, consciously or unconsciously, by the belief in reincarnation held by the Cathar heretics who at that time teemed in Italy and southern France, and who presumably had got it originally from India. It is significant that at just the same moment, about 1200, traces of metapsychosis are found also in western Judaism, in the Provencal *Cabbala*. But Francis held neither to transmigration of souls nor to pantheism. His view of nature and of man rested on a unique sort of pan-psychism of all things animate and inanimate, designed for the glorification of their transcendent Creator, who, in the ultimate gesture of cosmic humility, assumed flesh, lay helpless in a manger, and hung dying on a scaffold.

I am not suggesting that many contemporary Americans who are concerned about our ecologic crisis will be either able or willing to counsel with wolves or exhort birds. However, the present increasing disruption of the global environment is the product of a dynamic technology and science which were originating in the Western medieval world against which Saint Francis was rebelling in so original a way. Their growth cannot be understood historically apart from distinctive attitudes toward nature which are deeply grounded in Christian dogma. The fact that most people do no think of these attitudes as

Christian is irrelevant. No new set of basic values has been accepted in our society to displace those of Christianity. Hence we shall continue to have a worsening ecologic crisis until we reject the Christian axiom that nature has no reason for existence save to serve man.

The greatest spiritual revolutionary in western history, Saint Francis, proposed what he thought was an alternative Christian view of nature and man's relation to it: he tried to substitute the idea of the equality of all creatures, including man, for the idea of man's limitless rule of creation. He failed.

Both our present science and our present technology are so tinctured with orthodox Christian arrogance toward nature that no solution for our ecologic crisis can be expected from them alone. Since the roots of our trouble are so largely religious, whether we call it that or not, we must rethink and refeel our nature and destiny. The profoundly religious, but heretical, sense of the primitive Franciscans for the spiritual autonomy of all parts of nature may point a direction. I propose Francis as a patron saint for ecologists.

Substantial Healing

Francis Schaeffer

An essential part of a true philosophy is a correct understanding of the pattern and plan of creation as revealed by the God who made it. For instance, we must see that each step "higher"—the machine, the plant, the animal, and man—has the use of that which is lower than itself. We find that man calls upon and utilizes the animal, the plant, and the machine; the animal eats the plant. The plant utilizes the machine portion of the universe. Each thing, in God's creation, utilizes the thing that God has made under it.

We must also appreciate that each thing is limited by what it is. That is, a plant is limited by being a plant, but it is also limited by the properties of those things under it that it uses. So the plants can only use the chemicals on the basis of the boundary condition of the chemicals' properties. There is nothing else it can do.

But this is true also for man. We cannot make our own universe; we can only use what is under us in the order of creation. But there is a difference, and that is that the animal, for example, must use the lower *as what it is*. Man has to accept some necessary limitations of what is under him, but he can *consciously* act upon what is there. That is a real difference. The animal simply eats the plant. He cannot change its situation or properties. The man, on the other hand, has to accept some limitations, but nevertheless is called upon in his relationship to nature to treat the thing that is under him *consciously*, on the basis of what God has made it to be. The animal, the plant *must* do it; the man *should* do it. We are to use it, but we are not to use it as though it were nothing in itself.

Now let us look at it in another way. Man was given dominion over creation. This is true. But since the Fall man has exercised

138

this dominion wrongly. He is a rebel who has set himself at the center of the universe. By creation man has dominion; but as a fallen creature he has used that dominion wrongly. Because he is fallen, he exploits created things as though they were nothing in themselves, and as though he has an autonomous right to them.

Surely then, Christians, who have returned through the work of the Lord Jesus Christ to fellowship with God, and have a proper place of reference to the God who is there, should demonstrate a proper use of nature. We are to have dominion over it, but we are not going to use it as fallen man uses it. We are not going to act as though it were nothing in itself or as though we will do to nature everything we can do.

A parallel is man's dominion over woman. At the Fall—not before it, I think, but at it—man was given dominion in the home over the woman. But fallen man takes this and he turns it into tyranny and makes his wife a slave. So, first in the Judaistic teaching—the Old Testament law—and then later and more specifically in the New Testament, man is taught to exercise dominion without tyranny. The man is to be the head of the home, but the man is also to love his wife as Christ loves the church. Thus everything is back in its right place. There is to be order in the midst of a fallen world, but in love.

So man has dominion over nature, but he uses it wrongly. The Christian is called upon to exhibit this dominion, but exhibit it rightly: treating the thing as having value in itself, exercising dominion without being destructive. The church should always have taught and done this, but she has generally failed to do so, and we need to confess our failure. Francis Bacon understood this, and so have other Christians at different times, but by and large we must say that for a long, long time Christian teachers, including the best orthodox theologians, have shown a real poverty here.

As a parallel example, what would have happened if the church at the time of the Industrial Revolution had spoken out against the economic abuses which arose from it? This is not to suggest that the Industrial Revolution was wrong, or that capitalism as such is necessarily wrong, but that the church, at a point in history when it had the consensus, as it does not have now, failed (with some notable exceptions) to speak against the abuse of economic dominion. So also the church has not spoken out as it should have done throughout history against the abuse of nature.

But when the church puts belief into practice, in man *and in nature,* there is substantial healing. One of the first fruits of that healing is a new sense of beauty. The aesthetic values are not to be despised. God has made man with a sense of beauty, in a way no animal has: no animal has ever produced a work of art. Man as made in the image of God has an aesthetic quality, and as soon as he begins to deal with nature as he should—as having dominion but not exploiting nature as though it had no value in itself, and realizing it is also a creature of God as man is— beauty is preserved in nature. But also, economic and human value will accrue, for the problems of ecology that we have now will diminish.

Christians should be able to exhibit individually and corporately that, on the basis of the work of Christ, dealing with things according to the world view and basic philosophy of the Bible, they can produce something that the world has tried, but failed, to produce. The Christian community should be a living exhibition of the truth that in our present situation it is possible to have substantial sociological healings—healings that humanism longs for but has not been able to produce. Humanism is not wrong in its cry for sociological healing, but humanism is not producing it. And the same thing is true in regard to a substantial healing where nature is concerned.

So we find that when we begin to deal on a Christian basis, things begin to change; not just theoretical things, important as they are, but practical things. Man is not to be sacrificed, as pantheism sacrifices him, because, after all, he was made in the image of God, and given dominion. And yet nature is to be honored, each thing on its own level. In other words, there is a balance here. Man has dominion; he has a right by choice, because he is a moral creature, a right by choice to have dominion. But he is also by choice to exercise it rightly. He is to honor what God has made, up to the very highest level that he can honor it, without sacrificing man.

Christians, of all people, should not be the destroyers. We should treat nature with an overwhelming respect. We may cut down a tree to build a house, or to make a fire to keep the family warm. But we should not cut down the tree just to cut down the tree. We may, if necessary, bark the cork tree in order to have the use of the bark. But what we should not do is to bark the tree simply for the sake of doing so, and let it dry and stand there a dead skeleton in the wind. To do so is not to treat the tree with integrity. We have the right to rid our houses of ants; but what we have no right to do is to forget to honor the ant as God made it, out in the place where God made the ant to be. When we meet the ant on the sidewalk, we step over him. He is a creature, like ourselves; not made in the image of God, it is true, but equal with man as far as creation is concerned. The ant and the man are both creatures.

In this sense Francis' use of the term "brothers to the birds" is not only theologically correct, but a thing to be intellectually thought of and practically practiced. More, it is to be psychologically felt as I face the tree, the bird, the ant. If this was what "The Doors" meant when they spoke of "Our Fair Sister," it would have been beautiful. Why have orthodox, evangelical Christians produced no hymns putting such a beautiful concept in a proper theological setting?

One does not deface things simply to deface them. One would not willingly with no reason deface the rock. After all, the rock has a God-given right to be a rock as he made it. If you must move the rock in order to build the foundation of a house, then, by all means, move it. But on a walk in the woods do not strip the moss from it for no reason and leave it to lie by the side and die. Even the moss has a right to live. It is equal with man as a creature of God.

Hunting game is another example of the same principle. Killing of animals for food is one thing, but on the other hand they do not exist simply as things to be slaughtered. This is true of fishing, too. Many men fish and leave their victims to rot and stink. But what about the fish? Has he no rights—not to be romanticized as though he were a man—but real rights? On the one hand it is wrong to treat the fish as though it were a human baby; on the other hand, neither is it merely a chip of wood.

When we consider the tree, which is "below" the fish, we may chop it down, so long as we remember it is a tree, with its own value *as a tree.* It is not a zero. Some of our housing developments demonstrate the practical application of this. Bulldozers have gone in to flatten everything and clear the trees before the houses are begun. The end result is ugliness. It would have cost another thousand dollars to bulldoze *around* the trees, so they are simply bulldozed down without question. And then we wonder, looking at the result, how people can live there. It is less human in its barrenness, and even economically it is poorer as the top soil washes away. So when man breaks God's truth, in reality he suffers.

The hippies are right in their desire to be close to nature, even walking in bare feet in order to feel it. But they have no sufficient philosophy, and so it drifts into pantheism, and soon becomes ugly.

But Christians, who should understand the creation principle, have a reason for respecting nature, and when they do, it results in benefits to man. Let us be clear: it is not just a pragmatic attitude; there is a basis for it. We treat it with respect because God made it. When an orthodox, evangelical Christian mistreats or is insensible to nature, *at that point* he is more wrong than the hippie who has no real basis for his feeling for nature and yet senses that man and nature should have a relationship beyond that of spoiler and spoiled. You may, or may not, want to walk barefoot to feel close to nature, but *as a Christian* what relationship have you thought of and practiced toward nature as your fellow creature, over the last ten years?

Why do I have an emotional reaction toward the tree? For some abstract or pragmatic reason? Not at all. Secular man may say he cares for the tree because if he cuts it down his cities will not be able to breathe. But that is egoism, and egoism will produce ugliness, no matter how long it takes. On this basis technology will take another twist on the garrote of both nature and man. The tyranny of technology will grow to be almost total.

But the Christian stands in front of the tree, and has an emotional reaction toward it, because the tree has a real value in itself, being a creature made by God. I have this in common with the tree: we were made by God and not just cast up by chance.

Suddenly, then, we have real beauty. Life begins to breathe. The world begins to breathe as it never breathed before. We can love a man for his own sake, for we know who the man is—he is made in the image of God; and we can care for the animal, the tree, and even the machine portion of the universe, each thing in its own order—for we know it to be a fellow creature with ourselves, both made by the same God.

Nature, Humanity, and Biblical Theology Observations Toward a Relational Theology of Nature

Bruce C. Birch

First, we must look more closely at what the Scripture has to say about the *relationship of God to nature.* God's activity and concern are not limited to history nor is history to be considered a more important medium for divine activity. Gerhard von Rad, ironically the scholar most responsible for a *helisgeschichte* reading of the Old Testament, has written in an important essay:

> The greater part of what the Old Testament has to say about what we call Nature has simply never been considered. If I am right, we are nowadays in serious danger of looking at the theological problem of the Old Testament far too much from the onesided standpoint of an historically conditioned theology.

We can only scratch the surface of the great store of biblical materials to be explored here.

Nature is itself God's creation and has intrinsic worth before God apart from relationship to the human. An Israelite proverb claims, "The Lord has made everything for its purpose . . ." (Prov. 16:4). Israel at worship proclaimed "The earth is the Lord's and the fulness thereof, the world and those who dwell therein" (Ps. 24:1). Genesis 1:31 states unequivocally, "God saw *everything* that he had made, and behold, it was *very* good." Jesus claims that God values even the sparrow that falls.

Nature is itself a witness to God's work and his glory. "The heavens are telling the glory of God; and the firmament proclaims his handiwork," said the psalmist (Ps. 19:1). The Psalms and the Prophets are filled with such creation hymns witnessing to the evidence of God's glory, not in history but in nature.

Perhaps the greatest are Yahweh's speeches to Job out of the whirlwind (Job 38–41) where multitudes of creatures and natural phenomena (even "behemoth" and "leviathan") witness to God's creative power, and before whose witness Job seems insignificant. Jesus' stilling of the sea is nature's witness to Jesus' messiahship even before the disciples recognized him as the Messiah.

Nature itself can be an instrument or participant in God's activity in the world as in the Song of Deborah or Amos 4. Nature is called in as witness and judge in God's covenant lawsuit:

> Arise, plead your case before the mountains,
>   and let the hills hear your voice.
> Hear, you mountains, the controversy of the Lord,
>   and you enduring foundations of the earth. (Mic. 6:1–2)

Perhaps most importantly, nature has rights, and is as such the recipient of God's care. "One of the reasons for keeping the sabbath is so that the animals can rest" (Exod. 23:12), notes Fisher. In the laws of sabbath year and Jubilee the land itself is to be left fallow by divine commandment because it has the right of rest and renewal: "It shall be a year of solemn rest for the land" (Lev. 25:5).

Deuteronomy 20:19 reflects a perspective whereby elements of nature (here it is trees) have rights which cannot be subordinated to human need even if that need is considered urgent. "When you besiege a city for a long time, making war against it in order to take it, you shall not destroy its trees by wielding an axe against them; for you may eat of them, but you shall not cut them down. Are the trees in the field men that they should be besieged by you?"

The concept of the rights of nature already begins to lead us into a second area of concern for a relational theology of nature, namely *humanity's relationship to nature*.

Let me begin by acknowledging the understanding of human dominion over nature which appears in Genesis 1:28 and Psalm 8, and is reiterated in the New Testament by Paul. The language is royal language and this sovereignty of humanity is to be exercised as representative of God's divine sovereignty. It is not license for human indulgence. Exercise of dominion is accountable to God. Nevertheless, the tendency is to hierarchical thinking.

I believe that in order to understand this we have to grasp that for other Ancient Near Eastern cultures, especially Mesopotamian, humanity was dominated by nature which was identified with the gods and godesses of the pantheon. Human life existed only at the mercy of the divine powers, and the cult attempted to control those powers by magic and incantation· André Dumas writes helpfully on the work of Joël de Rosnay:

> De Rosnay speaks of the stage of survival, where prehistoric man, dominated by nature, wrested from it his subsistence. Then comes the stage of growth, where man the artisan dominates nature and exploits it without care of violating it. Today we are by the nature of things and by rational choice led to a third stage, that of conviviality with nature, where equilibrium should succeed growth as it succeeded survival. To further clarify de Rosnay one could say that prehistoric man surviving under nature is magical, the industrial man of growth over nature is analytical and that of man living in equilibrium with nature should be systemic.

Genesis 1 is important for its assertion of the created goodness of the whole natural order, and for its declaration of human freedom from the power of a divinely inhabited nature. It did set the stage for a view of humanity over nature which has been radicalized by subsequent generations of theology and culture in the now technologized West.

But the most important consideration for our exploration of the biblical tradition is that the Scripture also bears witness to the perspective of conviviality with nature that Dumas calls systemic. Dominion is not the only picture offered to us by biblical theology, and in fact, is hardly to be regarded as the dominant one (no pun intended).

There is ample witness in the Scripture to a view of humanity that regards us as limited with respect to nature, and at one with nature in our own standing before God. Humanity is within not above nature. If Psalm 8 asks "What is man?" and answers with dominion and status just below God, then we should also listen to Psalm 144 answer the same question: "O Lord, what is man that thou dost regard him . . . ? Man is like a breath, his days are like a passing shadow" (vv. 3–4).

Or listen to Psalm 49:12: "Man cannot abide in his pomp, he is like the beasts that perish;" or Isaiah 40:6–7: "All flesh is grass. . . . The grass withers. . . ." The Yahweh speeches of Job 38–41 are certainly no witness to human dominion over nature, "'Where were you when I laid the foundation of the earth?'"

In much of the Old Testament nature passages, humanity is pictured as a part of nature, a part of the created order under the sovereignty of God. God's intention was harmony within that creation characterized by the important word *shalom,* wholeness. We are related to the whole created order and called to seek *shalom* in it.

Genesis 2 attempts to stress this in its picture of humanity (*'adam*) taken from the soil (*'adamah*) with Eden as the symbolic picture of harmonious creation. I think that perhaps the most eloquent picture of the relationship of humanity to nature is to be found in Psalm 104. God is praised as Creator, and all sorts of things in creation are enumerated as witnesses to God's glory as Creator: the heavens, the seas, the valleys, the mountains, wild asses, cattle, trees, birds, goats, badgers, moon, sun, and lions. Then, verse 23, almost casually, states, "Man goes forth to his work and to his labor until the evening." The psalm then affirms that the whole earth "is full of thy creatures" (v. 25). Verses 27–30 say:

These all look to thee,
  to give them their food in due season.
When thou givest to them, they gather it up;
  when thou openest thy hand, they are filled with good things.
When thou hidest thy face, they are dismayed;
  when thou takest away their breath, they die
  and return to their dust.
When thou sendest forth thy Spirit, they are created;
  and thou renewest the face of the ground.

Many other passages could be cited as witness to this biblical view of harmony and interrelationship between humanity and nature.

This becomes extremely important when we begin to consider the nature of sin. It has been customary to speak of Genesis 3 as the story of the Fall. But this very term implies a tumbling down the hierarchical ladder to some lower level of being. In biblical terms sin is relational. Its consequence is the breaking of relationships, not a shift in the ontological character of human existence.

Look at Genesis 3. Humanity was created for whole relationship to God, others, and nature. The sinful, disobedient act creates a gap in relationship to God, erects barriers between humans one to the other, and alienates humanity from the very soil out of which human life was created.

Sin is not a matter internal to human life. It breaks the *shalom* intended by God in creation. Nature participates in the brokenness of sin, and Scripture clearly witnesses to this reality. The flood story is a tale of universal sin, but it is not so much a story of human nature, as one of the human and nature. God almost undoes the creation in the flood (it is not just a rain storm but the opening of the deeps, the waters of chaos). Its conclusion is the guarantee of the trustworthiness of nature.

The prophets speak of nature participating in the brokenness of sin. Hosea 4:1–3 announces God's lawsuit against Israel for covenant unfaithfulness and observes: "the land mourns, and all who dwell in it languish, and also the beasts of the field, and the birds of the air; and even the fish of the sea are taken away." Isa. 24:4–5 speaks in similar fashion of the consequence of sin, "The earth mourns and withers, the world languishes and withers; the heavens languish together with the earth. The earth lies polluted under its inhabitants."

In Romans 8, Paul says "the whole creation has been groaning in travail together until now" (v. 22). All of these passages speak of sin, not as a corruption of human nature, but as relational. It ruptures the full relationship of God, humanity, and nature, and all suffer the alienation which results.

But what of human power? Is our relationship to nature only one of destructive potential, or is there creative possibility as well? Here I can only share what seem to me are some helpful indicators of a biblical approach. It seems to me that Israel's understanding of election with respect to the nations has some carry-over in regard to the role of humanity within nature. Properly understood, Israel was called to missional responsibility but not to a special position of privilege among the nations before God. She was to live her life in order to be a blessing to all the families of the earth, and called to be a "light to the nations," but the prophets constantly reminded her that this did not mean God loved Israel more than the nations nor did this entitle them to special standing materially or spiritually.

When exclusivism reared its head new voices confronted it. The same understanding lies behind Paul's mission to the gentiles. In regard to nature it seems to me proper to see humanity commissioned to special responsibility with respect to nature, but warned that this is not indication of special love or privilege before God. The proper role is the role of servant. We are given responsibility but it is within the household of creation, not at its head. For this reason I like the commission of Genesis 2:15 which says literally that humanity is placed in the Garden of Eden (symbolic of created wholeness and harmony) "to serve it and to keep it (or care for it)." An interesting contrast is that in Ancient Near Eastern myths of creation, humanity is created as the servants of the gods, but in Genesis 2 humanity is the servant of the soil and its produce.

Finally let us turn to a third area of concern for a relational theology of nature. I wish to speak of *redemption as new creation*. Is the redemptive activity of God and the redemptive work of Christ directed and limited to the human? Many would say so, but I hardly see how this is possible biblically. Unless we see human life lived in a vacuum, redemption must involve nature as well because redemption is precisely God's work to restore the relationships that have been broken by sin. This is why new creation is such a helpful image of God's redemptive work.

New creation cannot be defined apart from God's original creation. Thus, the wholeness of relationship which God intended must also be the goal of his redemptive activity, the redemption wrought in Jesus Christ and the redemptive task of the church. Nothing less than *shalom* can characterize the relationship of humanity and nature which the church must seek.

New creation first appears as a theme in the Old Testament, most fully in Deutero-Isaiah. Here the salvation history and creation history are wedded. They are both a part of the work of one God. God's redemption of Israel and the nations renews nature as well.

> For you shall go out in joy,
>   and be led forth in peace (*shalom*);
> the mountains and the hills before you
>   shall break forth into singing,
> and all the trees of the field shall clap their hands.
> Instead of the thorn shall come up the cypress;
>   instead of the brier shall come up the myrtle.
>   (Isa. 55:12–13)

In the New Testament I believe that redemption as new creation is also relational. The concern is not for some ontological restoration of human existence, a mere repair job on one of the broken orders of creation. New creation points to the restoration of wholeness in relationship for all of creation. Jesus himself exemplified this wholeness not only in relationship to persons but in his attitude toward all creation. He pointed easily to nature for examples of God's care ("Consider the lilies of the field . . ."). He spoke to the stormy sea and it became still. He walked on its surface. These are not indicators of human dominion. They point to the harmony of the divine, the human, and the natural in Christ. In Paul, new creation also relates to nature. Just as nature "groaned in travail" from sin, so too, in Christ as new creation "the creation itself will be set free from its bondage to decay . . ." (Rom. 8:21).

Is nature an integral part of the biblical picture of redemption? The final yes must come from all those glorious eschatological pictures of the future of God's kingdom in its ultimate fulfillment, in the fulness of time. They do not picture some disembodied spiritual union with the godhead for that higher order of humanity. They paint glorious pictures of God, humanity, and *nature* in rich communion. I have only room to cite one of the most eloquent of these pictures of nature as eschatological witness.

> The wolf shall dwell with the lamb,
>   and the leopard shall lie down with the kid,
> and the calf and the lion and the fatling together,
>   and a little child shall lead them.
> The cow and the bear shall feed;
>   their young shall lie down together;
>   and the lion shall eat straw like the ox.
> The sucking child shall play over the hole of the asp,
>   and the weaned child shall put his hand on the adder's den.
> They shall not hurt or destroy
>   in all my holy mountain;
> for the earth shall be full of the knowledge of the Lord
> as the waters cover the sea.
>
> (Isa. 11:6–9)

I wish to close by suggesting that *shalom* is an appropriate biblical symbol for a relational theology of nature. *Shalom* is a remarkably powerful Old Testament concept which needs fuller exploration. Its basic meaning is wholeness, and it is God's will for creation. In it are united the spiritual and the material; the natural and the human; the Creator and the created. It is a powerfully universal symbol. And yet, *shalom* is also used to express the full life of the particular. *Shalom* is the well-being of any creature, any thing when it is fulfilling its created purpose.

In the New Testament, Ephesians 2:13–14 reads, "But now in Christ Jesus you who once were far off have been brought near in the blood of Christ. For he is our peace, who has made us both one, and has broken down the dividing wall of hostility. . . ." This might just as well be read "He is our shalom," and I wish to suggest to you that in the spirit of *shalom* it is time for us to consider whether the far who has been brought near might not be our nonhuman neighbor as well as our human one, and whether the dividing wall of hostility which has been broken down might not also be that barrier which has for so long separated humanity and nature.

Theocentrism
The Cornerstone of Christian Ecology

Vincent Rossi

To be Christian is to be ecologist. This statement, which I believe to be absolutely fundamental to an authentic Christian understanding of the relationship between Man and Nature, flatly contradicts the generally accepted opinion among secular environmentalists that the Christian religion and Christian doctrine are somehow responsible for the present, anthropocentric paradigm which is so destructive to the environment. The truth is exactly the opposite. Contrary to the opinions of Lynn White and Arnold Toynbee (to name two of the more prominent exponents of the anti-Christian position), it was only when western civilization "liberated" itself from Christianity and its traditional doctrinal restraints that the door was opened for the ecological disasters of the present day. A truly impartial reading of history makes this fact so clear that one wonders at the persistent and pervasive hostility so many environmentalists display toward Christianity. The main point of White's essay has been so easily refuted by a number of Christian writers that it is more than curious that he continues to this day to be quoted with approval in environmental publications. One would almost think that many people want to use Christianity as a convenient scapegoat.

Be that as it may, the problem goes deeper than mere hostility or a simple prejudicial reading of history. The problem lies in the phenomenon of *modernity*. Modernity is not just a particular "paradigm" or its accompanying mind set. Nor should any criticism of modernity be automatically written off as nostalgic yearning for a "simpler" past. We are dealing with an unprecedented psychological and spiritual rupture of a people from its cultural and spiritual roots. The roots of western culture are Judeo-Christian: the radical reversal of this historical hierarchy of values in the West resulting in the Renaissance, the Reformation, the so-called Enlightenment, and the industrialization and secularization of society, produced the modern world as we know it.

All of these results were anti-Christian in effect; that is, they tended to reduce the influence of Christianity on the prevailing social order. Modernity has produced rationalism, materialism, rampant individualism, secularism, relativism, humanism, technism—all "heresies," from a Christian perspective, that is, perversions or reversals of traditional Christian doctrine and practice.

Further consequences have been the loss of the *traditional orthodox spirituality* originally suited to the western soul, and especially significant for us today, the loss of a *sense of continuity* with our own past. As a people we have no sense of the history of our *consciousness.* All of these things came about not as a result of, but at the expense of Christianity.

The so-called Protestant ethic was not an authentically Christian development in spirit or substance. It was a secularized distortion of genuine Christian doctrine. Max Weber, the writer who made so much of the *concept* of the Protestant ethic, mistakenly identified the essence of Christianity as the *source* of this secularized distortion. He, too, was a product of the very civilization that stole the stones from the pyramid of authentic Christian doctrine, ethic, and praxis to build the foundation of a civilization that was radically anti-Christian, antitraditional, and antinature in effect. We are all products of that civilization. Its influence on our souls and psyches goes perhaps deeper than many of us would care to admit.

Perhaps this helps to explain the inability of many otherwise perceptive ecological theorists to distinguish the authentic Christian "ecosophy"—which is biblical, patristic, and in principle supportive of the values of true ecology—from the anthropocentric domination and exploitation of nature supported by the so-called Protestant ethic which was only a conventionalized, rationalized caricature of true Christianity.

Authentic Christianity has been subjected to an unprecedented critical attack, both from within and without. No other culture in history has subjected its traditional, perennial religion to such attack with such weapons (the full arsenal of modern, rationalistic, critical analysis) as has done the modern West. This resulted in subverting the values of traditional Christianity and replacing those values, rooted as they were in the restraints— and the vision—of Christian asceticism and mystical contemplation, with a set of values that were originally derived from the priceless heritage of Christianity, but denuded of all spiritual purpose and meaning—in a word, secularized.

The modern western paradigm is related to Christianity only as a *heresy*—a subtle counterfeit, a demonic caricature of authentic Christian values. Such is the "established religion" of modernity, the "opiate" of the new, secularized, industrial civilization. The anthropocentric paradigm is not Christian, not even close. It is a *new faith entirely.* And if some Christian denominations have tended to capitulate before the new faith, they do so not as biblical and traditional Christians but as moderns who are nominally Christian but who are as lost and rootless as the rest of the moderns.

If ecology can be called the "subversive science" because it undermines all the values of the present paradigm, authentic orthodox Christianity is even more subversive. Unlike current ecological thinkers who naively look to form a new paradigm or consensus based on ecological principles married to economic and political realities, Christianity—if it is true to its biblical, patristic, and traditional roots—cannot align itself ultimately with any paradigm, for "My kingdom is not of this world."

Of course there is no escaping the fact that the materialistic, evolutionistic, scientistic, anthropocentric paradigm of modernity did arise in the West, within the traditional social framework of Christianity. But it should be clear by now that the phenomenon of modernity is not specifically a Christian or a *western* phenomenon. It is as antiwestern as it is antieastern, as much anti-Christian as it is anti-Buddhist, as well as antifolk-culture, antinature, etc. Eastern cultures, as we have seen, are equally susceptible to its blandishments and encroachments. In fact, it may be the West that first wakes up from the hypnotic dream of modernity and takes up the responsibility to begin casting off its destructive values.

However, the way we in the West begin to attack the anthropocentric paradigm of modernity should not be to call, naively, for the creation of a "new paradigm." This would be as though it were actually within our power to alter the very roots of a culture's subconscious presuppositions, the "supersubconscious" of the western soul, so to say.

Rather, we should seek to understand something about the history of consciousness (not the same as the academic "history of ideas" or "intellectual history") and the consciousness of history. We need to return to the place where the continuity of our *normal* folk-culture and spiritual tradition was ruptured. It is at that place that we will find the Grail where God, Man, and Nature may be seen in their true harmony. I believe that it is there that our words and work on behalf of deep ecology will acquire the power to effect a change in consciousness.

It is ironic that the very attempt to forge a "new paradigm" so popular among some ecophilosophers seems to be an impulse that is of the same spirit as the paradigm of modernity they seek to replace. Modernity is, after all, based on just the kind of worship of the new, the belief in evolutionary progress, and the faith in human reason that seem to be the basic presuppositions of ecophilosophers. It matters not that the evolutionary progress is transposed from the technological to the "spiritual" plane, or that the "revolution" called for is on the level of concepts rather than politics. It is still the spirit of modernity at work.

What is needed to counter the paradigm of modernity is not a "new paradigm" but the perennially true paradigm (which is not, by the way, the sole property of any single system of thought or belief).

The concept of the "perennial philosophy" is often evoked by ecophilosophers as a possible source for deep ecological principles. Few seem to realize that, even if the concept of the perennial philosophy as commonly understood is valid, which is open to question, its principles are diametrically opposed to the attempt to forge a new philosophy or a new paradigm by the syncretistic piecing together of all the elements in modern western philosophy and eastern religion that seem to support the notion of biocentrism.

If the *sophia perennis* holds a key for deep ecology and ecophilosophy, it is not in the worn-out futilities of Heideggerian existentialism or in the pantheism of Spinoza (to name two of the most quoted names) that one will find it. Only in the living and perennial church—the church of all times and places—does the *sophia perennis* actually exist *as such:* as the perennial wisdom of her truest representatives, the holy fathers and mothers, saints, sages and ascetics of the Christian tradition. In the patristic tradition, *paradigma* means *exemplar;* and exemplar means an interpenetrating image by which God creates and sustains all things (Gen. 1:27). From the patristic perspective, then, there are no "new paradigms"; there are only true (or false) paradigms.

It seems to me that the weakest aspect of deep ecology theorizing is the religious/philosophic. This is understandable since the movement is in its infancy and has not yet found its philosophical roots. But there does seem to be a tendency toward an absence of *distinguo,* and a lack of intellectual rigor in ecological philosophizing. For example, to call for a "new metaphysics," as we see often in the literature, betrays a failure to understand (in the same manner discussed above regarding paradigms) that metaphysics, if capable of being grounded in truth or adequate (*adaequatio*) to reality, cannot ever be "new" but must be grounded in first principles that are *perennially* true.

There may be new expressions of metaphysical truth, but the metaphysical dimension itself cannot change or be changed. This is one of the basic insights of the so-called "perennial philosophy." The "strict data" for a metaphysical science preexists in pure Intellect, or Logos. Such objective metaphysical "data" can only be "extracted" by means of a real and efficacious spirituality, which in Christianity has found its highest expression in the patristic, hesychastic, Palamite tradition.* All the rest is mere reasoning. No matter how brilliant, it remains blind.

Another example of loose thinking among non-Christian ecophilosophers is to make glib reference to the so-called "Christian/Cartesian" man-nature dualism. In fact, René Descartes represents precisely the rupture with the principles of Christian philosophy that begins the modern era of philosophy. Cartesian dualism, though it treats of Spirit and matter, God and nature, decisively separates the two "realms" in a way that is radically anti-Christian. Cartesianism is not a product

---

* St. Gregory Palamas, a fourteenth-century monk from Mt. Athos, brilliantly defended the validity of hesychastic spirituality as developed and transmitted by the early Desert Fathers, against theological influences originating in the Roman Catholic Church.

of Christian thought; it is the result of a rejection of the principles of Christian philosophy. But all *modern* philosophers are children of Descartes, even those who labor most mightily to restore the lost unity, such as Heidegger, whose atheistic presuppositions render null and void his otherwise brilliant philosophic labors. His "let beings be" is not so much a deep ecological insight as it is an admission of failure to illuminate logically the ontological opacity of Being in beings. Heidegger attempts to accomplish what the fathers of the church have already done, but without their faith, the indispensable means of extracting the necessary metaphysical "data." Hence Heidegger's nullity. He can make nothing of "I am that I am." So also the *Gelassenheit* of Heidegger can have neither the meaning nor the result of Meister Eckhart's *Gelassenheit* of which it is only an atheistic and individualistic counterfeit.

This brings up the matter of dominion and stewardship. The biblical term "dominion," as had been so often pointed out by Christian writers, does not mean and never has meant *domination* of nature by "Lord Man." It is time ecophilosophers let Christianity off that hook. The biblical concept of dominion is connected to two other key ideas—*covenant* and *stewardship.* The Bible expresses not only God's covenant with humanity, but, more importantly, God's prior covenant with all of nature. Dominion implies the responsibility to *serve* nature, or God in nature.

This leads us to stewardship. Christian ecological thinking has always emphasized stewardship. But in light of the widespread use of that term to refer to the secular notion of the management of natural resources, I have almost become convinced of the necessity to abandon the notion of stewardship as too compromised. The biblical idea of stewardship has become identified, although incorrectly, with the "wise management" anthropocentrism of the Pinchot-inspired° resource conservation and development movement of modern technocratic society. Many ecologists seem to feel that stewardship is a notion that is not *deep* enough to meet the needs of the environmental crisis, or that Christian stewardship supports a biologically unwarranted *chauvinism* of human beings toward all other species and even the earth herself.

° Gifford Pinchot, early twentieth-century conservationist, known as President Theodore Roosevelt's "chief forester."

This attack on stewardship, though well intentioned and certainly understandable, may be counterproductive in the long run. The biblical concept of stewardship, if correctly and completely understood in all its social implications and spiritual amplitude, remains (despite my earlier reservations) a valid and adequate foundation on which to build a deep ecology ethos. In fact, in the West, where our spiritual roots remain Judeo-Christian despite several centuries of secularized rationalism, the biblical concept of stewardship may be the only effective platform for a deep ecology movement.

Those western ecologists who look to oriental religion for concepts (however valid in the eastern religious and psychological sphere) in support of an already abstract and rationalistic notion of biocentrism may ultimately be doing the deep ecology movement more harm than good, because they are ignoring or denying our western ethical and psychological roots which, whether we like it or not, happen to be Judeo-Christian. What good is an elegant biocentrist environmental platform buttressed by Buddhist or Taoist philosophy, if it satisfies certain philosophers but fails to move the hearts and minds of the masses of people who make the conscious and unconscious decisions that cumulatively create the environmental crisis?

This brings us to the anthropocentrism/biocentrism dichotomy and debate. From the Christian point of view this is actually a false dichotomy and it is leading the debate down a blind alley. There is another alternative to anthropocentrism, and it is precisely orthodox Christianity that may show us the way, if we will allow it.

A Christian must be opposed to anthropocentrism in all its forms. This may sound surprising in view of the widespread assumption among ecologists that Christianity and anthropocentrism are identical. But anthropocentrism is only the result of the cult of individualism, which is one of the basic assumptions of modernity, and which Christianity has opposed throughout its history. This statement can be supported by such an abundance of ascetic and mystical texts from every era of Christianity that we need not belabor the point.

As a Christian, and as someone who seeks to think clearly in accord with universal metaphysical principles, I am also opposed to an unqualified acceptance of biocentrism, for the simple reason that there is an inescapable contradiction at its heart, namely, absolute biological egalitarianism. One simply cannot claim that humankind is just another species and in the same breath talk of "holistic" or "righteous management." Mankind is not just another species, if only because no other species has need of the concept of "righteous management," or even, for that matter, of "biocentrism" in order to maintain its "place" in nature.

While fully accepting the fact that humankind is inextricably interrelated (the basic premise of ecology) with all other species and beings in Nature, we must also admit that there is an irreducible quality of absoluteness to the human state, which is related to our self-awareness and our freedom, and which quality, at the same time, is not *caused* by the biological interrelatedness itself. Certainly modern humanity needs to be humbled, and anthropocentrism needs to be denounced as a pernicious cancer upon the earth. Biocentrist thinkers are to be commended for so doing; but absolute biocentrism, in the sense that it denies the spiritual centrality of the human state, must be avoided, for despite its sense of moral superiority over anthropocentrism, it is not grounded in the principles of metaphysical, cosmological, and spiritual knowledge, but in a sentimental and rationalistic reaction. However much we may sympathize with that reaction, we cannot compromise immutable metaphysical principles. To do so would be to lose all hope of arriving at the genuine metaphysical and ontological intuitions that will give birth to the renewed and resacralized cosmology that is so vital to an effective ethic of the environment.

The traditional Christian patristic answer to anthropocentrism is not biocentrism but *theocentrism*. The basic insight of theocentrism is: The earth is the Lord's. I said "insight" rather than "premise" because the idea that "the earth is the Lord's" is not a philosophical concept but an *intuition* flowing from *faith*. It is an intuition that illumines the earth and all species with intelligibility, beauty, and glory.

Theocentrism preserves the Christian insight into the nature of humankind as the image of God, without giving humanity license to dominate or despoil the earth, for Gaia belongs to the Lord who is a consuming fire and a just Judge, as well as a loving Father who has married himself to our Mother Earth with an everlasting covenant (Gen. 9:9–13).

At the same time, theocentrism embraces the ecological insight of the radical interrelatedness of all species, *homo sapiens* included. Christian theocentrism is in perfect agreement with a statement Bill Devall, a noted ecophilosopher, made in his paper, "The Development of Natural Resources and the Integrity of Nature," page 11:

Insofar as we are distinguishable as individual organisms, the insight is that all other individuals and species also have an equal "right" to their home on the planet and to develop their own form of realization.

The Christian theocentric position would also accept the Four Laws of ecology as stated by Barry Commoner: 1) everything is connected to everything else, 2) everything must go somewhere, 3) nature knows best, and 4) there is no such thing as a free lunch.

Thus from the standpoint of patristic Christian theocentrism, everything is naturally connected to everything else because God is All in all and everything is connected to God, who *is* the first principle of theocentrism. Everything must go somewhere, because there is no "thing" without a "where"; and thingness and whereness are the matter (*mater:* Mother) of existence, which is the mystery of "standing out from" Being or God. Nature knows best because the infinite wisdom of Divinity is embodied in creation, or as Augustine put it, "all creatures contain the trace of the Trinity." (One of the names of God in scholastic theology is *Natura naturans;* of course, Nature—as God's handiwork—knows best!) There is no free lunch, because interrelatedness explains the ecological meaning of the proverb "as you sow, so shall you reap."

For an orthodox Christian, anthropocentrism is clearly a false and untenable position, while biocentrism could be accepted only on condition that human beings are not defined solely by their biological place in nature. Biocentric egalitarianism is a valid insight when dealing with humans as "individual organisms."

But when dealing with human beings in their wholeness as *images of God*, biocentrism will be seen, by Christians at least, as a half-truth, for it is in our unique *personhood* that the image of God rests; the mystery of the Person transcends biology. And yet, as all nature is a symbol of God, that is, as all creatures reflect God's wisdom, beauty, and unity as living symbols, then biocentric interrelatedness must be a reflection in the physical world of an interrelatedness of a higher order.

This insight we see expressed in Christian patristic theology as the fact that we only find our true personhood in the Person of Christ, because "in him, we live, and move, and have our being" (Acts 17:28). Therefore, for the Christian, theocentrism more perfectly expresses the Christian's perspective in the deep ecology movement.

Theocentrism condemns the tragic distortions of anthropocentrism, while affirming mankind's priestly role at the center of creation. Theocentrism turns stewardship away from management, wise or unwise, and toward *servanthood*. Theocentrism affirms biocentric egalitarianism on the level of individual species and beings, while underscoring the undeniable uniqueness of the human person as temple of the Holy Spirit and image of God.

As priest, Man celebrates nature, rediscovers the mystery and sacredness in our Mother Earth, and worships God in and through nature. As servants, human beings do not presume to know best, but seek to learn and study from nature to see how to help insure the "self-realization" of all creatures, mostly by learning how to get out of the way and let the wonderful intelligence embodied in creatures be released from human encroachment.

In sum, the Christian theocentric position sees Man as celebrating servant; earth as Mother "married" to God the Father by everlasting covenant; all creatures, great and small, as children of God; and the cosmos as a vast eucharistic liturgy in which all things have their beginning and their end in Christ.

Without an authentic Christian-based expression of an ethic of the environment, I believe the deep ecology movement, no matter how important its insights, will not succeed. This theocentric environmental ethic can be called the Eleventh Commandment: "The Earth is the Lord's, and the fullness thereof; thou shalt not destroy the earth nor despoil the life thereon." Any position short of a full acceptance of theocentrism will not represent the completely biblical, patristic Christian worldview. From the standpoint of the authentic Christian hierarchy of values, which does not change with the times, it remains perennially true that to be Christian is to be ecologist.

"All Creation Groans"
Theology/Ecology in St. Paul

James A. Rimbach

Introduction

For theologians to address the issues of the ecology and environment is a typical "test case" for doing theology today. It is typical for several reasons. In the first place, the world's populace has been much seized with this issue for about twenty-five years. Man and his environment in the last quarter-century has seldom been discussed as a dispassionate issue: more often as a "crisis issue." It is not at all uncommon for theologians to take up such popular concerns following the lapse of at least a decade or two; indeed, fifty years might prove closer to the average. (This phenomenon is known in theological circles as "being in the forefront" and "taking a prophetic stance!") With some popular issues the attempt to avoid fadism results only in last-minute bandwagoning; that is, fadism after the fad is over. But the environment is a topic that is no fad and will not go away. Environment and ecology will be a dominant concern of the foreseen and the unforeseen, possibly much more frightening, future.

Secondly, this particular issue demonstrates that in many instances, theologians, without the lead of the secular world, have difficulty finding matters of interest and relevance to discuss, except for esoteric and "in-house" matters.

Thirdly, whenever topics are taken up in theological circles by way of response to secular discussions, there is a great temptation for the practitioners of theologies of the right to bless the *status quo* and to say to the powers that be: "Do as thou hast done, and the Lord be with thee."

Fourthly, theology by response presents a temptation also to those of theologies of the left completely to rewrite Scripture and tradition in such a way as to discover the true and eternal theological base for the latest radical programmes—the more radical, the more blessed.

And finally by way of introduction, I must wonder if theological reflection on the ecological issue in the third-world will be dominated by the seductive slogans that sometimes permeate, but by no means advance, third-world discussions on many matters. The reality of the emerging global village that is the world of the late twentieth century and will certainly be that of the twenty-first leads me to be wary if I am told that ecological/ environmental crises are: a) imported from the West, or b) exploitative tricks, or c) malevolent plans hatched by multinational corporations. The problem with these notions is simple, but not always easily discernible—namely, that they can hinder rather than help the people of Asia and other areas constructively to address the issues at hand, and that is what they must do.

I would risk the opinion that, whatever special shapes the ecological crises of the world take in different areas, cultures, countries—in almost every case the proximate cause can be reduced to the underlying pressures of rapidly increasing and increasingly urbanized populations. This is particularly evident in Asia, on the subcontinent and Africa; more so than in many other areas of the world.

The sins and excesses which nature was in the past able to absorb with little visible effect have simply reached beyond threshold proportions and we cannot dismiss our various bad habits in the way that we have done previously, with or without theological approbation.

I think that we must also bear in mind that human depredations of the environment which quite equal and sometimes exceed those of the developed world and places under its influence can be located in areas outside the interest and influence of the developed countries, the multinationals, colonialists, imperialists, and assorted other "bad guys." As far as "mother earth" is concerned, the dangerous predator is the human race, period.

The Christian theologian's resource of Holy Scripture contains a considerable amount of awareness of, reflection upon, and analysis of, the relationship of created people and created world. It is one particular instance of this resource that will be the subject of examination here: the eighth chapter of St. Paul's Letter to the Romans and its twin themes of "faith-hope-renewal" and "disobedience-despair-decay."

The pairing and contrasting of the two thematic groups "disobedience-despair-decay" and "faith-hope-renewal" is a constant feature of the biblical message. They represent the extremes of human experience that stand on both sides of the drama of the Cross. What we hope to elaborate here is the manner in which these terms embrace not only humankind, but express the experience of the entire created order.

The biblical drama is not written in terms that *separate* the destiny of man and the rest of creation, but is always related in such a way as to stress a common destiny, a linkage of humanity and nature which is symbiotic. The spiritual destiny of humankind has "repercussions" throughout the world of nature. To understand this emphasis of the scriptural message one must realize that what happens to one part of creation accompanied with rationality and self-consciousness is experienced by the other irrationally and without self-consciousness, neither being more real or, for that matter, more important. The most elaborate expression of this commonality in the Scriptures is a treatment of St. Paul in his Letter to the Romans, chapter eight, which we reproduce here under the heading:

Subjected in Hope

I consider the sufferings of the present to be as nothing compared with the glory to be revealed in us. Indeed the whole created world eagerly awaits the revelation of the sons of God. [*hē gar apokaradokia tēs ktiseōs tēn spokalypsin tōn huiōn tou theon apekdechetai*] Creation was made subject to futility [*tē gar mataiotēti hē ktisis hypetagē*] not of its own accord but by him who once subjected it; yet not without hope, because the world itself will be freed from its slavery to corruption and share in the glorious freedom of the children of God. Yes, we know that all creation groans and is in agony even until now. [*oidamen gar hoti pasa hē ktisis sustenazei kai synōdinei achri tou nyn*] Not only that, but we ourselves, although we have the spirit as first fruits, groan inwardly while we await the redemption of our bodies. In hope we were saved. But hope is not hope if its object is seen: how is it possible for one to hope for what he sees? And hoping for what we cannot see means awaiting it with patient endurance.

The Spirit too helps us in our weakness, for we do not know how to pray as we ought; but the Spirit himself makes intercession for us with groanings that cannot be expressed in speech. He who searches hearts knows what the Spirit means, for the Spirit intercedes for the saints as God himself wills. (Rom. 8:18–27 NAB°)

This relatively neglected excerpt from the apostle Paul requires a close examination and will yield a good deal of understanding, especially for Christians seeking guidance and a sense of responsibility in matters of ecological concern and the relationship of humankind to the natural world. It will, perhaps, be most helpful to follow the argument in outline form, and then return to consider the details.

SUBJECTED IN HOPE—DESTINED FOR GLORY

NOW: Theme A: There is no glory without suffering. ("No Cross—No Crown!")
This is true of the entire creation.

Theme B: God's creation has been "subjected";
and—
This subjection was made by God.
The *cause* is human revolt (sin).

Theme C: The subjection was made "in hope," that is, in view of a coming reversal/renewal.
The renewal in nature, like its subjection, is linked to human spiritual destiny.

THEN: (Theme C) The "hope" in which creation has been subjected will be realized with the revelation of the full inheritance, the "sonship" of Christians.

(Themes B–C) The creation will share in the freedom that is now gradually unfolding. This freedom is not a human achievement, but a gifting of God's Spirit.

° NAB stands for New American Bible.

NOW: (Theme C) We have a "first-fruit" experience, together with a longing for completion, both of which indicate the reality of our hope.

(Theme A) This process can be compared with "birth-pangs"; that is, a temporary suffering that produces ultimate joy.

(Theme C) In view of what is to come, patience is in order, and a confidence, because our hope is directed to and controlled by something outside of ourselves.

NOW AND THEN COMBINED (Theme A) The experience of the Spirit testifies to the assurance of this developing reality: The Spirit is the downpayment of what is to follow.

(Theme C) The resultant Christian hope for all the created order is indomitable and confident, because it is Spirit-led.

Good Beginnings

Now we shall return to the beginning of the Apostle's argument and follow through the various themes and their development in more detail. Underlying the entire presentation is the simple assumption that God's world is a whole.

Of course, it is natural to view this from the human perspective—indeed, we have no other possible point of view. But even though our viewpoint is, of necessity, anthropocentric, it does not need to be anthropocentric in a subjective manner, that is, by reducing everything in the cosmos to the values that we discern there for our own benefit narrowly defined. Our proper place in creation is determined by our recognition that our position and relationships are something which have their purpose and value from God.

Since we know God as the Creator of everything, we cannot devalue anything in nature simply because we, as yet, have not discerned its value to selective human endeavors. We want to view all things with which we share the earth, and the earth itself, as entities which find their highest purpose in praising and pleasing not man, but God. This is a great challenge to faith, because the ground of such praise and pleasure to God will not always be apparent.

God had, according to the biblical creation account, a good deal of practice in making the pronouncement, "How good it is!" *before* he came to speak the pronouncement of approval on his work of creating humankind. The universe, light, seas, vegetation, animals: of all of these he spoke, "How good it is!" (Gen. 1). And all interrelated, interdependent, and doing fine, prior to the introduction of the humans. It is a serious usurpation of authority for us to view any part of this creation with the decision: "not necessary," or "not good," "not important," or "fit for extinction." This right has nowhere been delegated by God. He maintains reservation of judgment for himself.

The Subjection of Nature

When St. Paul asserted the subjection of nature, he undoubtedly had in his mind's eye the Genesis account of the "fall" (Gen. 3).

Cursed is the ground because of you (v. 17).

In the scriptural elaboration of the curse there is depicted a new relationship between humans and nature. The humans are expelled from the "garden" (actually an orchard) and set on a way of obtaining vegetation for food as a result of hard labor. Now they are surrounded, not by fruit-bearing trees, but expanses bearing thorns and thistles. Bread will be obtained by the sweat of the brow in an endless cycle of sowing and reaping; a cycle from which one is delivered finally and only by returning to the ground from which he came. Human sin pollutes the earth and drags an innocent nature into the circle of futility which human rebellion has established. This "effect of the curse" should not be understood as a haunting that issues only from primeval times. The statement of Genesis is theological and paradigmatic. In Deuteronomic theology, the blessing and/or curse involving the natural environment is attendant upon covenant fidelity or infidelity, as the case may be (see Deut. 28). The pragmatic statements of covenant theology in Deuteronomy derive their impetus from the prophetic tradition of Israel, and a case in point is the expression of Isaiah 24:4–6:

The earth mourns and withers,
the world languishes and withers;
the heavens languish together with the earth.
The earth lies polluted under its inhabitants;
For they have transgressed the laws,
violated the statutes,
broken the everlasting covenant.
Therefore a curse devours the earth,
and its inhabitants suffer for their guilt;
therefore the inhabitants of the earth are scorched,
and few men are left.

It is an equally arresting thought to take account of God's grace as portrayed in the subsequent Genesis account of the great flood, the third great scene of the biblical primeval drama. There, nonhuman life has far greater representation in the ark than the solitary family of Noah which represents the human race. Nor is the selection depicted in any way as having a utilitarian basis in relation to the humans. The Creator is concerned that there be a full representation of his glorious creation in its new beginning. None is thought too small—or too big—or expendable.

Even those which might be thought to be exempt from Noah's salvage effort, animals designated *unclean* (i.e., not fit for human consumption) are included: "to keep their kind alive upon the face of the earth" (Gen. 8:17). At the conclusion of the account, the voice of the Lord vows *"never again to curse nature because of man"* (Gen. 8:21). The postdeluge blessing is *first* uttered over the animals—*then* imparted to Noah (Gen. 8:19–22; 9:1). But the *subjection* remains and continues:

The fear of you and the dread of you shall be upon every beast of the earth, and upon every bird of the air, upon everything that creeps on the ground and all the fish of the sea; into your hand they are delivered (Gen. 9:2).

Oddly enough, this statement of the heavy responsibility that humans bear in relation to other living things has been taken by many people throughout long periods of history as a license to exploit and despoil the environment with impunity! And yet the gracious covenant is with "every living creature!" (Gen. 9:15–17). No wonder the created order finds itself in *agony* and *futility,* as St. Paul asserts. It has witnessed the human distort its mandate for creative stewardship into a program of defoliation and depopulation that at times borders on the demonic. Such depredations are occasionally called "taming the wilderness" or "sport." The theologian might well opine that this kind of double-speak was learned, not from the Creator of the garden, but from its primeval forked-tongue denizen!

Nor will the anthropologist and psychologist pass over the prominence of masculine, sport-dominance-aggressiveness metaphor that often comes into play here. As has been said in quite another context: if you don't know the difference between rape and romance, you've got a problem!

There is yet another connection to be made between the natural world and the destiny of the human. Not only is the "subjection" of nature linked to the rebellion of humankind, but also the character of the subjection: futility (*mataiotēte*) is linked to the fact that the full disclosure of the saints as "sons of God" waits its obvious appearance. Subhuman or nonhuman creation is frustrated in its attempts to fulfil the purpose of its existence since God has so destined that without humankind it should not be made perfect;

> We may think of the whole magnificent theatre of the universe together with all its splendid properties and all the chorus of sub-human life, created to glorify God but unable to do so fully, so long as man the chief actor in the drama of God's praise fails to contribute his rational part.[1]

The same sentiment is expressed by Karl Barth, who writes:

> The occasion of the dislocation and longing and *vanity*, presented to us in the whole creation, is not this or that particular pain or abomination or absence of beauty, not even the sum of observable disadvantages attaching to the world as we see it: the occasion is rather createdness itself, the manifest lack of direct life, the unsatisfied hope of resurrection. We cannot, surely, pronounce the created world to be direct, genuine, and eternal life. The perpetual interaction of energy and matter, of coming into being and passing to corruption, of organization and decomposition, of thirst for life and the necessity of death: this *bondage of corruption* which encompasses all living creatures from the microbe to the Ichthyosaurus, and from the Ichthyosaurus to the

---

1. C. E. B. Cranfield, *Romans* [International Critical Commentary] J. A. Emerton, C. E. B. Cranfield, eds. (Edinburgh: T. & T. Clark, 1975), 414.

most distinguished professor of Theology—everything, that is to say, which we know as "life" or, by the analogy of anything that we comprehend as life, can conceive as such, this surely is not direct, genuine, and eternal life.[2]

It is interesting to note that commentators on Paul all find it necessary to establish whether or not St. Paul is exhibiting an interest in nature "for its own sake" or, as some finally declare, only in relation to the lesson which the saints are supposed to derive from its agonized subjection.[3] Apropos of that discussion, one should hear Luther's comment: "Most exegetes take the term 'creature' in this passage to mean man, . . . But it is better to understand man through the word 'vanity.'!"[4]

Hope in Subjection

The subjection to futility of the natural world is not such that it does not continue to give its witness to the Creator. That much is true for Paul, certainly, when he writes (Rom. 1:20), "Ever since the creation of the world his invisible nature, namely, his eternal power and deity, has been clearly perceived in the things that have been made." It was true also of Jesus, whose summons to "consider the lilies" and to "look at the birds" served to illustrate his most profound spiritual challenge, that to ultimate trust. "By looking for liberation from the forces which rule it, . . . it (i.e., creation) manifests the hope in the midst of hopelessness which makes sense only to faith."[5]

Other commentators also have given eloquent expression to the character of hope in the creation. E.g., Karl Barth:

Beyond pessimism and optimism, where the origin of the vanity of the COSMOS in the unobservable Fall of the creature from the Creator is apprehended—there emerges hope, hope of the restoration of the unobservable union between the Creator and

2. K. Barth, *The Epistle to the Romans* (trans. from the sixth edition by E. C. Hoskyns). London: Oxford, 1960 (1933), 308.

3. See Cranfield, *Romans,* p. 415 and note 2. Also E. Käsemann, *Commentary on Romans* (trans. and ed. by G. W. Bromiley). Grand Rapids: Eerdmans, 1980, 232.

4. M. Luther, *Lectures on Romans,* Wm. Pauck, ed. [Library of Christian Classical] Philadelphia: Westminster, 1961, 237.

5. Käsemann, *Romans,* 236.

the creature, through the Cross and Resurrection of Christ. Once freedom is seen.[6]

And we add these comments from E. Käsemann's commentary:

> The world makes sense as creation only if it is oriented constitutively to Christian liberty, which is not to be equated with automony. . . . This is to be seen as the pledge also of its participation in eschatological liberation. If Marcion was forced by the inner logic of his theology to cut out vv. 18–22, he is followed today by an existentialism which individualizes salvation and thereby truncates Paul's message by describing freedom formally as openness to the future. In fact, it is a term for the earthly reality of Christ's lordship. . . . The truth in the existential interpretation is that it recognizes in pride and despair the powers which most deeply enslave mankind. Its theological reduction derives from a world view which no longer knows what to do with Pauline apocalyptic, allows anthropological historicity to conceal the world's history, obscures the antithesis of the aeons of [Romans] 1:20ff, by natural theology and here through the assertion of mythology, and for this reason can no longer speak adequately of the dominion of Christ in its worldwide dimension.[7]

Properly to appreciate Paul's themes, we must see that both the "subjection" of nature and its "hope" have dimensions beyond the romantic. There is more to the futility of creation than its endless cyclic repetition of flowering and decay, its "redness of tooth and claw" (Lipsius), its beauty and catastrophe, and its mindless exploitation.

The curse of subjection can also be seen in the fickleness of humankind's ever-changing and ill-considered dispositions toward animal life: the neurotic affection for kept animals on the part of some; baiting, torture, and painful slaughter; the penchant for zoos of the sort that would manifestly dispirit any living thing; the callous cruelty of some experimental techniques. All of these too are part of the "folly" in regard to the created order and the "folly" experienced by it (Rom. 1:23).

Neurosis and indeed pathology are evidence of the fallenness of the steward of creation when animals are accorded care

---

6. Barth, *Romans,* 308.
7. Käsemann, *Romans,* 236.

and concern *in the place of* necessary charity to fellow humans, and when the less-than-human is accorded human (and more!) qualities: pampered, spoiled, worshipped and finally venerated in pet cemeteries. This too qualifies with the Apostle's strictures about "exchanging the glory of the immortal God for images resembling mortal man or birds or animals or reptiles" (Rom 1:23). It does not give the pattern for responsible Christian relationships to the natural world.

Christians will discern the foolishness and the futility in some currently popular projections for a new relationship with nature that call for a "new animism." Here too there is displacement (and often an accompanying attempt at discrediting) of the Creator through the folly of those who elevate the created to an illegitimate place: a new idolatry. But such essentially idolatrous and pagan notions will not bear the fruit that their advocates expect. There is no "Spirit" behind such programmes, which substitute style for theology. And hope—humankind and for the natural world—is the especial precinct of the Spirit; that is the argument of St. Paul in Romans 8. *The creation waits with eager longing for the revealing of the sons of God.*

> . . . not that creation will share the same liberty-resulting-from-glory as the children of God will enjoy, but that it will have its own proper liberty as a result of the glorification of the children of God . . . the liberty proper to the creation is indeed the possession of its own proper glory—that is, of the freedom fully and perfectly to fulfil its Creator's purpose for it, that freedom which it does not have so long as man, its lord (Gen. 1:26, 28; Ps. 8:6) is in disgrace.[8]

It is just at this point that the theological investigation of Romans 8 becomes particularly problematic. St. Paul utilized two particular metaphors to describe the transition of the created order from bondage to freedom: to wit, *agony* and *travail.* These metaphors are commonly cited as the indicators of the painful passage from one world to the next, as indeed they sometimes, but not always are. Because of this association, the entire discussion by St. Paul is sometimes cast in the shadowy conceptions of the "eschatological." The metaphors of agony

---

8. Cranfield, *Romans*, 416.

and travail have their home in the Old Testament, and are often linked to messianic expectations in their subsequent use.[9]

In the Old Testament itself, this imagery is a quite general and natural one used to express the thought of severe distress which nonetheless has a happy outcome. Its occasional use in (proto-) apocalyptic contexts does not make them *ipso facto* eschatological or apocalyptic.

In commentary on Romans the metaphor of travail has been linked with the likewise common expectation of renewal and transformation of Israel which is often said, both in the Old Testament and in the Apocryphal literature to be attended by the renewal and transformation of nature. This is by no means totally out of place, but it should be noted that the connection is not explicit in Romans 8.[10]

Thus, a rather artificially constructed "eschatological" format results. Since Genesis 3 presents a "Paradise Lost" which is today read in largely nonhistorical if not mythological terms, the counterbalance of restoration and renewal—"Paradise Regained"—is termed an "eschatological" or at least a "parahistorical" motif. If we try to fathom the "why" of this, part of the answer may be because in this manner theologians are able to continue to talk about a redemption that is largely invisible to them because they insist on defining it in sociopolitical terms. This, oddly enough, approximates the hermeneutical situation in the New Testament portrait of rabbinical Judaism against which the New Testament writers struggle.

Pauline eschatology is certainly a legitimate subject for investigation. But in Romans 8, and in the context of the explication of the linkage of man and nature under curse and promise, we are dealing with *Christology,* not Messianism. I cite as my text for this assertion none other than Käsemann (one who otherwise holds the term eschatological in special fondness):

> Grace relates us more deeply to the earthly because it thrusts the community as a whole and each of its members beneath the cross

---

9. Is. 26:17; 66:8, *et al.;* Jer. 4:31; Micah 4:9. In the New Testament, Mark 13:8; John 16:21; 1 Thess. 5:3. For discussion see S. Mowinckel, *He That Cometh* (New York: Abingdon, 1954), 261–279.

10. Jubilees 1:29; 4:26; 1 Enoch 45:4f.; 72:1; 91:16f.; 2 Esdras 7:75; 2 Baruch 3:7–4:1; 31:5–32:6; 44:12; 51:3; 57:2; 73:6–74:4; IQS 4:25; IQH 11:13–14; 13:11–13.

where extreme assault and victory coincide. In opposition to the enthusiasts Paul had to go back to Jewish apocalyptic to present it thus. The widespread dislike for this tradition is a chief obstacle in interpreting the text, since avoiding Scylla inevitably means being caught on Charybdis. The Gentile-Christian enthusiasm for a radically realized eschatology has finally gained the upper hand in church orthodoxy too, and in so doing has obscured the theology of the cross.[11]

So we repeat the assertion that Paul's argument is Christological and not Messianic, and for two reasons: 1) the literary traditions that he cites are used to connect the renewal of nature associated with the *parousia* of the Son of Man, and not a preparatory event to a "Messianic" age, and 2) the hope of which he speaks is not the usual "Messianic" (i.e., today, "chiliastic") hope, but one that is seen only through the Cross.

Lest this all seem petty cavil with obscure terminology, we can illustrate what differences result from hermeneutical points of view with two contrasting statements.

A) J. A. Baker, "Biblical Attitudes to Nature"
Christianity's . . . answer to the problem . . . significantly enough, is not a means of redeeming the world of nature as well as the soul of man, so that they can then live in harmony to create the Kingdom of God on earth, but a spiritual liberation of those men and women who believe in Jesus, who must then wait in patience for a total remaking of the cosmos in God's own time, and by God's own hand. The book of Revelation, with its vision of a new heaven and a new earth (Revelation 21:1), is the logical culmination of this approach.[12]

B) Eric C. Rust in Nature—Garden or Desert?
Since the redemption is already being wrought out within the historical scene, the renewal of man's relationship to nature is one dimension of the Church's redemptive mission. The salvation of the individual, with eyes mainly upon a transcendent and heavenly order, ignores the fact that the setting of this universe is straining forward expectantly for the final unveiling of the sons of God. But, as already stated, the Christ makes such sonship available to historical man. Hence,

---

11. Käsemann, *Romans*, 232.
12. J. A. Baker, "Biblical Attitudes to Nature," in *Man and Nature*, Hugh Montefiore, ed. (London: Collins, 1975), 107.

it must be accompanied by the preparation of man's natural habitat for that final unveiling.[13]

The hope of which St. Paul speaks is not a hope which renders people *inactive*, but a hope which puts them into action. Here the Old Testament background might prove helpful: in Hebrew the word for wait and hope are the same (*qāwāh*), but the emphasis is not on "patience" or "inactivity" but upon ultimate trust which accompanies a program of action.

When Cranfield writes:

The full manifestation of our adoption is identical with the final resurrection of our bodies at the Parousia, our complete and final liberation from the (futility) and (decay) to which we—like the subhuman creation—have been subjected.[14]

The words can be read as hopeful or as a counsel of despair. The basic hermeneutical issue seems to be the imprecision with which biblical commentators confront poetic imagery. This is complicated for this writer by the fact that the (American) Lutheran traditions have no developed tradition of eschatology. Cranfield, for example, goes on to note that the prophetic utterances of renewal for Israel and for humankind in general are frequently accompanied with expressions of joyous praise by the created order (see Is. 11; 40–66 *passim* and Jewish extrabiblical literature).[15] And this is indeed the case. But this prophetic, and poetic, imagery is not a programatic eschatology. It is simply a literary impress that grows out of the unitary view of the Bible on man and nature. So it is that all biblical hopes for the future involve the imagery of natural renewal.

It is to be emphasized that there is not "one" prophetic hope nor one so-called "eschatological" picture.[16] St. Paul's, to me, does not seem to be Isaiah's, nor, for that matter, does Paul's seem to be that of Jewish apocalyptic writers. Paul's vision of an unfolding freedom for the cosmos is not the disclosure of apocalyptic (or eschatological) blueprints, but is born from his confidence

---

13. Eric C. Rust, *Nature—Garden or Desert?* (Waco, Tex.: Word, 1971), 33.
14. Cranfield, *Romans*, 419.
15. See the references in notes 9 and 10.
16. The oft-cited references in 2 Esdras 7:62ff., and 10:9ff., for example, I find to be of quite a different direction than the development of Paul's argument. Still less compatible is IQH 3:7ff.

that *in Christ all the promises of God find their Yes!* (II Cor. 1:20). Paul has transformed, as did Jesus, the idea of the Kingdom, and that brought with it certain adaptations of the attendant and hoped-for renewal of the natural order.

A Happy Ending

In summary, St. Paul asserts in Romans 8 that *to the degree* (perhaps even a small one) that the saints have the Spirit as a pledge, to that small degree the realization of the hope of created things is also glimpsed. It is not at all clear whether it advances understanding to call this eschatology, much less "realized" eschatology. Nor is it satisfactory to speak here of some inexorable movement toward a final accomplishment, unless we remember that both the pledge and the goal are the Spirit's transforming work and not the achievement of enlightened individuals or groups. It seems that a more enlightening approach would be to connect the "revealing of the sons of God" in Romans 8 with the restoration of the divine image which allows man to function in his intended and primevally entrusted stewardship role in regard to the natural world, and which therefore also allows the natural world to that degree to enjoy the fulness of its being and enjoy the promised freedom from its subjection to vanity because of humankind.

If I read Karl Barth correctly (and of this one is never sure), he agrees. He concludes the pertinent portion of his remarks on the subject with a quotation from Nietzsche:

When the sons of God shall appear, "because of their appearing, Nature, which never leaps, makes one leap, and that a leap for joy; for then it knows that at last and for the first time it has attained its goal."[17]

I believe that here Barth (and Nietzsche) have caught the genius of St. Paul. Here is no "lamb lying down with the lion" in the enjoyment of millennial bliss and indolence and for the entertainment of regenerate and unregenerate man. But instead we have here a glance at the world around us through glasses marked with a Cross. The first "eye-opening" event for the cross-bearing Christian is to see the great chorus of the

---

17. K. Barth, *Romans*, 309.

groaning creation with whom he now feels a new affinity and responsibility.

To be sure the groaning of that blessed order is matched by our own groaning, even as the Spirit helps us with our first prayers. And as we see in Christ our means of release from the bondage to corruption as the result of our rebellion, we all together give our attention to the promise: we resume our "image" status in relation to the creation, not so much to "transform" but to realize, to reach our purpose as creatures in company with all other created things.

Unfortunately, to my mind, Barth continues on to cast all of this as thoroughly "eschatological"; i.e., as if groaning together were a sufficient goal. But, can we not object: hope that has no realization, however partial, is not hope either, but dreams, fantasy (apocalyptic eschatology!)? To be sure, popular theology often embraces both fantasy and eschatology in one breath and with great enthusiasm. This is but an over-existentialization of the Christian faith: it feels great but bears no scrutiny.

I conclude with one suggestion and two quotations. First the suggestion that the correct manner in which to gain understanding of the connection between Old Testament biblical prophetic nature-imagery and New Testament eschatological nature-imagery in relation to the divine economy of redemption is to observe the redeemed of the Kingdom exercising their Spirit-led stewardship of God's garden, whether or not there are theologians among them.

The first of the quotations is from Martin H. Franzman (emphasis mine):

> . . . all creation groans. But since Christ came and God pronounced anew his benediction on the creature man and on man's creature world, the world's agony is the agony of travail; there is in it the promise of a glad new birth. . . .

> Paul teaches us to hope for a re-creation of the world. . . . (God) would not have sent his Son into the world *if he were minded to take us out of this substantial world,* as disembodied spirits, into some vague and insubstantial heaven of His own. *The ministry of Jesus was God's yes to his creation spelled out in act;* . . . The promise of our resurrection is a promise of the resurrection of the body. We wait for a heaven and an earth that shall be a wondrously and unimaginably *new* heaven and earth. *But the same continuity that makes the body of the future one with our present*

body connects the new unsullied world of God with the world we know, the world whose frustrate beauty makes us marvel still, whose futile workings can testify to him who once said "Very good!" and will again say "Very good!" to all His hands have made. [18]

And finally Luther:

He says two things: First, the creature will be delivered, namely, from vanity, when the wicked will have been damned and taken away and the old man abolished. Such deliverance happens even now every day in the saints. [19]

---

18. Martin H. Franzmann, *Romans* [Concordia Commentary]. St. Louis: Concordia, 1968, 149–150.
19. M. Luther, *Lectures on Romans*, 238–239.

God's Joyous Valuing of Nature

H. Paul Santmire

As joy, the creative rule of God gives value to all things and gives the world its goodness. This theme comes into clearer focus when we recall the concrete image of God as the master builder. We can say that the master builder *takes delight* in what he has established and shaped. This, as we have seen, seems to be the significance of the Priestly writer's repeated expression in the creation narrative, "And God saw that it was good." A similar note is struck by the psalmist in his creation-hymn: "May the Lord rejoice in his works" (Ps. 104:31b). It is also rehearsed in the wisdom literature: God's creative shaping of the world, his wisdom, "beside him like a master workman," is "daily his delight" (Pr. 8:30).

But it is not immediately clear that the theme lends itself to translation into the language of value-theory. Is it meaningful to say that when God rejoices in his works he is giving value to the world? It *is* meaningful, so long as we remember that we are here concerned first of all with the creative rule of *God* and only in a secondary sense with the world. Our question, in other words, is not what is valuable within the world but what is valuable for God. We look not to what man may or may not value, but to the ultimate value-giver, to the "center of value."

The introduction of the technical language of value at this point will not sound so strange when we consider that the master builder, when he rejoices in the work of his hands, at the same time makes the judgment that what he has done is done *well*, that he has built at least as well as he had expected to build. So it is said: "And God saw that it was good."

Still, the central question remains unanswered: In what sense does God rejoice in all his works? In what sense does he value all things? Or to put it in slightly different form: What is the meaning of the goodness of creation?

It is instructive to note that for Karl Barth, the goodness of the created realm is its status as the external ground for the covenant of Grace. Creation is good because it is an instrument which makes possible the actualization of reconciliation and redemption. God rejoices in all things because he sees all things strictly as the preparation for the coming of Grace into the world:

> There is surely a realm of Nature [here Barth means the whole realm of creation, man together with heaven and earth] which as such is different from the realm of Grace. But at the same time in the realm of Nature *all* properties which are not directed to Grace and which do not come from Grace amount to nothing. *There is nothing in Nature which may lead a life for itself and which may follow its own course.*

From the perspective of Barth's theology, the master builder must be pictured as rejoicing in his establishing and shaping only because of something *still to be added* to his work, something for which his work is the occasion or the means.

The biblical writers, however, though they conceive of creation as a stage posited by God for the sake of redemption, do not conceive it *merely* as a stage. This is indicated most succinctly perhaps by the psalmist's hymn which celebrates God's creative works and his rejoicing in them with virtually no indication that creation is a stage for redemption (Ps. 104; cf. Ps. 29).

The theology of the Bible suggests that the created realm *has value in itself for God.* Creation, to use Jonathan Edwards's terminology, is presented as an *ultimate end,* not a subordinate one. The master builder is depicted, in other words, as rejoicing in his establishing *as such* and in his shaping *as such.* He values each epoch of his creative work in itself, not just as a stage on the way to the new creation.

In this respect Jonathan Edwards presents a view of creation considerably more adequate to the biblical view than Barth's. Edwards distinguishes between "original and independent ultimate ends" and "consequential and dependent ultimate ends." He then applies this distinction to the doctrine of creation:

> We must suppose that God, *before* he created the world, had some good in view as a consequence of the world's existence that was

originally agreeable to him to bring the universe into existence in such a manner as he created it. But *after* the world was created, and such and such intelligent creatures actually had existence, in such and such circumstances, then a wise, just regulation of them was agreeable to God, *in itself* considered.

The created realm, then, is a consequential and dependent ultimate end in which as such God rejoices.

Having considered the meaning of the divine rejoicing with respect to the relation between creation and the new creation, we now must consider the same theme with respect to the created realm itself. Here the specific problem is the value of the natural world. Does the Creator rejoice in his works in nature *only because nature is a stage* which God posits for the sake of man, or does God rejoice in his work in nature *also because nature is for him an ultimate end?* There is little doubt that the biblical teaching, and therefore the proper theological teaching, is that nature *is* posited by God for the sake of man, that God does rejoice in nature as a means for furthering his purpose with man. But is this the *only* reason why God rejoices?

Our earlier exegetical discussion has produced the clear answer that God rejoices in his work in nature as an ultimate end. God values it in itself. Recall two biblical references to animals of the deep. After God created the "great sea monsters," the Priestly author tells us in his familiar phrase that "God saw that it was good" (Gen. 1:21). That is, God delights in the sea monsters as such. He wills to have a realm in which there are such beasts for his own enjoyment.

In a similar way, the psalmist writes that God created the Leviathan in order to play with it (Ps. 104:26). God establishes and shapes a natural being without reference to the being and well-being of man. God wills to have the Leviathan for himself.

When we thrill to contemplate the almost incomprehensible reaches of our universe and try to understand that our world is more millions of years old than we can imagine, when we confront the infinitely vast spaces through which astoundingly large galaxies move or consider even the story of the evolution of life on our own planet, we may well believe that the majestic King of this universe rejoices that he has done and is doing all these works for *our* sakes. For the Bible describes a God who relates himself in a special way to man and who gives the world its being so that that special relation might be established.

Yet we may *also* believe, with the biblical writers, that all this temporal and spatial vastness is no mere stage or instrument posited solely for us. We may believe that the whole of nature is of intrinsic worth to God who rejoices in all his works as such. He calls the stars and planets by name (Is. 40:26; Ps. 147:7; Job 38:37), hears the cries of the animals in the wilderness (Ps. 104:27; 145:16), feeds the birds of the air (Mt. 6:26) and notices when the least sparrow falls to the ground (Lk. 12:6). He clothes the field with lilies (Mt. 6:30), and sports with the cosmic Leviathans. He wills to have the galaxies and the ichthyosaurs and the alligators and the lilies and the infinite seas of electrons, mesons, photons, and all other particles and forces for their own sakes quite apart from what they all may or may not mean for our being and well-being, as well as for the sake of our being and well-being.

In this way the King of the universe prepares all things for their transcendent Day of fulfillment, the new creation. Throughout a sequence of cosmic epochs he rules majestically in, with, and under his whole created realm. He powerfully establishes all things and so gives being and becoming to nature. He wisely shapes all things and so gives order and beauty to nature. He joyfully values all things and so gives goodness to nature. When the time comes that he judges that his finally fulfilled Omega-world can be ushered in, then he will bring his work in the present creation to an end. The creative rule of God will be consummated in the ultimate manifestation of the Divine Kingdom. Then the wolf will lie down with the lamb, and death will be no more. All things will be gloriously transformed when the day of the new heavens and a new earth dawns. The integrity of nature, its life with God, its intrinsic value in his eyes, the whole divine history with nature, will have been vindicated.

The North American Conference on Christianity and Ecology

Implementation Document

Section I

A Prayer of Thanksgiving and Confession

For the marvelous grace of Your Creation—
We pour out our thanks to You, our God,
    for sun and moon and stars,
    for rain and dew and winds,
    for winter cold and summer heat.
We pour forth our praise to You
    for mountains and hills,
    for springs and valleys,
    for rivers and seas.
We praise You, O Lord,
    for plants growing in earth and water,
    for life inhabiting lakes and seas,
    for life creeping in soils and land,
    for creatures living in wetlands and waters,
    for life flying above earth and sea,
    for beasts dwelling in woods and fields.
How many and wonderful are Your works, our God!
    In wisdom You have made them all! (cf. Psalm 147–148; 104:24)
But we confess, dear Lord,
    as creatures privileged with the care and keeping of Your Creation,
    that we have abused your Creation gifts
    through arrogance, ignorance, and greed.
We confess risking permanent damage to Your handiwork;
    we confess impoverishing Creation's ability to bring You praise.
Yet, we confess that Your handiwork displays Your glory,
    leaving all of us without excuse
    but to know You. (Romans 1:20)
We confess that Your handiwork
    provides the context of our living;

it is our home,
it is the realm in which we live the life of Your kingdom:
Your kingdom that is now in our midst and coming yet more fully.

We confess, Lord, that we often are unaware
of how deeply we have hurt Your good earth
and its marvelous gifts.
We confess that we often are unaware
of how our abuse of Creation
has also been an abuse of ourselves.

O Lord, how long will it take before we awaken to what we have done?
How many waters must we pollute?
How many woodlots must we destroy?
How many forests must we despoil?
How much soil must we erode and poison, O Lord?
How much of Earth's atmosphere must we contaminate?
How many species must we abuse and extinguish?
How many people must we degrade and kill with toxic wastes
before we learn to love and respect Your Creation;
before we learn to love and respect our home?

For our wrongs, Lord, we ask forgiveness.

In sorrow for what we have done
we offer our repentance.
We pray that our actions toward You and Your Creation
are worthy of our repentance;
that we will so act here on earth
that heaven will not be a shock to us.

May Thy kingdom come and Thy will be done on earth!

In our hearts, Lord, we promise anew
to reverence Creation as a convicting witness
of Your presence,
eternal power, and
divine majesty. (cf. Psalm 19 and Romans 1:20)

We promise to reverence Your Creation as a gracious gift
entrusted to us by You, our God.
We promise anew to be stewards
and not pillagers
of what You have entrusted to us.

Indeed, we acknowledge
that You have entrusted Your very Self to us
in Creation, through your Incarnation in Christ Jesus.
For the Word through Whom all Creation was made
and is upheld,
that Word "became flesh and dwelt among us,
full of grace and truth." (John 1:14)
Lord, you pitched Your divine tent among us
and made our home Your home as well.

Creator God,

You have given us every reason
to learn and promote the wisdom of lives
lived in harmony with Creation.

May we, your servants, increasingly serve.

May we, your servants, increasingly come to love Your Creation
as we also increasingly come to love You,

through Christ Jesus,
our Lord. Amen.

Section II

A Theology of Creation and Redemption

The whole of scripture presents God's activity in Creation and redemption; it describes the response and responsibility of human beings to the Creator, to each other, and to the rest of Creation.

The book of Genesis gives account of the Creation, with humanity as its climax. "God created the human being in his own image . . . male and female he created them. God blessed them and said: 'Be fruitful and multiply, and replenish the earth and subdue it'" (Genesis 1:27–28). Subduing and having dominion must be understood within God's covenant with humanity and Creation; its proper understanding must result in a loving care and keeping of the Creation in cooperative harmony with the Creator. The Creator whose "compassion is over all that he has made" (Psalm 145:9) put the human creature in "the garden . . . to till it and keep it" (Genesis 2:15).

God's covenant with Noah was made, also, with "every living creature . . . for all future generations" (Genesis 9:12). God's covenant with the people of Israel followed their deliverance from bondage and included the gift of "a good and broad land" (Exodus 3:8), together with the call, ". . . choose life . . . loving the Lord your God, obeying his voice" (Deuteronomy 30:19–20).

In their stewardship of the Creation, people are called to sustain the integrity of the land and of its creatures, to permit Creation's restorative and sustaining processes to work as the Lord wills. In their caring and keeping of the Creation, human beings are responsible for providing people, animals, and the land their times of rest, peace and restoration; their solemn sabbaths before the Lord (cf. Exodus 20:8–11; 23:10–12; Leviticus 25, 26; Luke 4:16–22).

The Psalmist reminds us of the Lord's dominion over Creation and of the integrity required of those who serve its Creator:

> The earth is the Lord's and its fulness,
> the world and those who dwell in it . . .
> Who shall ascend the mountain of the Lord?
> And who shall stand in his holy place?
> He who has clean hands and a pure heart . . . (Psalm 24:1–4).

Regrettably, throughout history human beings have been unfaithful to God's covenant; they have failed to exercise their vocation as stewards responsibly. The consequences for humanity and all of Creation have been devastating.

The prophets called people to repentance for being unjust, for holding idols in their hearts, for muddying clear waters, for trampling good pastures (cf. Ezekiel 14:1–6; 34:18–19); the prophets called people to renewed fidelity to the Lord and the Lord's covenant. But the covenant is broken and all Creation suffers the consequences of human faithlessness. Yet, hope remains for renewal of the covenant, for redemption both of humanity and the rest of Creation:

> I will make for you a covenant on that day with the beasts of the field, the birds of the air and the creeping things of the ground; and I will abolish the bow, the sword, and war from the land and I will make you to lie down in peace . . . (Hosea 2:18–20).

Building upon this hope and stressing that humanity must be freed from slavery to sin and selfishness, Jesus of Nazareth called people to the new life in God's kingdom, in communion with the living God, and in community with one another. He condemned the idolatry of serving self or mammon (money), urging his followers to seek first the kingdom of God and God's righteousness (Matthew 6:33). Righteousness means right relations with God, with one another, and with all creatures.

Dominion must be exercised as service in imitation of our Lord Jesus Christ who came not to be served but to serve and to give his life as a ransom for many (Mark 10:45). Jesus is our model: "Do nothing from selfishness and conceit . . ." writes the Apostle Paul. "Have this mind among yourselves, which was in Christ Jesus . . . who emptied himself, taking the form of a servant. . . . He became obedient unto death. . . . Therefore God has exalted him . . ." (Philippians 2:3–11).

We stand in need of redemption from the terrible effects

that greed, selfishness and various forms of idolatry have inflicted upon the earth, the waters, the atmosphere and all life. The Apostle Paul, in his Letter to the Romans, offers a vision of hope for restoration: "Creation itself will be set free from bondage to decay and obtain the glorious liberty of the children of God . . ." (8:21). This is accomplished in the redemptive work of Jesus, in his death and resurrection. Through him all things are reconciled with God (Colossians 1:20).

The book of Revelation also points to the prophetic picture of a New Creation. This is the vision of the new heaven and new earth toward which God is working in, through, and with us, until the consummation of history: "Behold I make all things new!" (Revelation 21:5).

Section III

The Ethics of Faith and Action

We are called to live out our faithfulness to God the Creator and Redeemer Who is revealed to us in the Creation (Psalm 19; Romans 1:20), in the Scriptures (2 Timothy 3:14–17), and through the testimony of the Spirit (Romans 6:16). Our living in faithfulness to the Lord is always in a particular historical and ecological context, a context within which God continues to act.

We live in a time of unprecedented awareness of the majesty of Creation. The universe reveals the Creator in a spectacularly convincing way. The astounding variety and integrity of life reflects God's perfection. And through us, God's human creatures, has come a self-awareness of the Creation itself; through us the magnificence of the Creation can be perceived. It also is through us human creatures, as bearers of God's image, that we also have our important role of loving, protecting, and cooperating with the Creation, and in so doing to better and more fully worship our Lord and Creator.

While we live in this time of unprecedented awareness, we also live in a time of unprecedented crisis within the Creation. The whole earth and all of its people are threatened with destruction. We confess that this results from human failure to discharge faithfully the responsibility of stewardship.

Many Christians and others are now hearing clearly God's call for a return to our proper role within the created order. We are being called to responsibility for the care and keeping of the

Creation and for sharing equitably the gifts and fruits that God provides.

As we respond to the Lord in this time of peril and promise, we believe that God calls us to:

1) a personal metanoia—a change of heart, renewed again and again—whereby we become sensitive and responsive to the presence and activity of God in nature and in one another and thereby also to the virtues of faithful stewardship.

2) the preservation and keeping of Creation's rich and diverse life, the healing of injuries inflicted by human arrogance and carelessness and greed, and the restoring of Creation's integrity.

3) the transition to an ecologically sustainable economy in which all members of the human family participate and from which they receive sufficient sustenance;

4) the renewal of an ethic of work as a blessing to self, others, and the Creation;

5) the transformation of the prevailing self-centered and overly consumptive lifestyle toward a lifestyle based on conservation, restoration, simplicity, shalom and abundant community;

6) the vigorous pursuit of justice in our communities, countries and world to overcome oppression and despoliation of people and nature;

7) the practice of unceasing prayer and persistent practice, directed at bringing all humanity into active service in the restoration and keeping of the Creation, through Christ Jesus.

Notes

Chapter 3 Why We Are in This Mess

1. John McPhee, "The Control of Nature," *The New Yorker*, February 23, 1987, p. 40. I have relied on McPhee's excellent article for the background concerning the Mississippi and Atchafalaya Rivers, and the confrontation with the Army Corps of Engineers.
2. Ibid., 49.
3. Ibid., 84.
4. Ibid., 52.
5. Ibid., 85.
6. Ibid., 84.
7. Ibid., 49.
8. Quoted in Lis Harris, "Brother Sun, Sister Moon," *The New Yorker*, April 27, 1987, p. 90. For a fuller discussion of the Lynn White, Jr., article, and a critique of it, see my book *A Worldly Spirituality* (New York: Harper and Row, 1984), 31ff.

9. Quoted in James M. Gustafson, *Ethics from a Theocentric Perspective* (Chicago: University of Chicago Press, 1981), 158.

10. Ibid., 272–3.

11. *The State of India's Environment, 1984–5* (New Delhi: Centre for Science and Environment, 1985), 209.

12. Brad Lemley, "The Soft Path of Amory Lovins," *The Washington Post Magazine*, June 29, 1986.

13. Barry Commoner, "A Reporter at Large (The Environment)," *The New Yorker*, June 15, 1987, 56.

14. Ibid., 57.

Chapter 4 Biblical Wisdom

1. This quote is from Thomas Berry's paper, titled "Christianity and Ecology," which was written as a critique of the proposed statement prepared for this conference.

2. For a bibliography containing many examples of such books and articles, see my book, *A Worldly Spirituality*, pp. 204–07.

3. Robert K. Johnston, "Wisdom Literature and Its Contribution to a Biblical Environmental Ethic," in *Tending the Garden*, Wesley Granberg-Michaelson, ed. (Grand Rapids: Eerdmans, 1987), 69.

4. For a full discussion of these biblical passages, see *A Worldly Spirituality*, pp. 59–65.

5. Ernst Käsemann, *New Testament Questions of Today* (Philadelphia: Fortress Press, 1969), 180. See also Käsemann's *Commentary on Romans* (Grand Rapids: Eerdmans, 1980).

6. Translation is by Paulos Mar Gregorios and appears in his essay, "New Testament Foundations for Understanding the Creation," *Tending the Garden*, 84.

Chapter 5 Hopeful Signs

1. The books are: Edwin R. Squiers, ed., *The Environmental Crisis: The Ethical Dilemma* (Mancelona, Mich.: AuSable Trails Institute of Environmental Studies, 1982), and

Wesley Granberg-Michaelson, ed., *Tending the Garden*. A full list of papers available from AuSable can be obtained by writing to them at: Route 2, Big Twin Lake, Mancelona, MI 49659.

2. Ted T. Cable, "Environmental Education at Christian Colleges," *Perspectives on Science and Christian Faith* 39, No. 3 (September 1987), 165–168.
3. David Douglas, "The Spirit of Wilderness and the Religious Community," *Sierra*, May/June 1983, 56.
4. Ibid., 57.
5. Diane E. Sherwood and Kristin Franklin, "Ecology and the Church: Theology and Action," *The Christian Century*, May 13, 1987, 473.
6. Gus Polman, "Developing Links to Share in Creation," *Earthkeeping* 3, No. 2, 1987, 3.

Chapter 6   Cooperative Technology

1. Albert Borgmann, "Prospects for the Theology of Technology," in Carl Mitcham and Jim Grote, ed., *Theology and Technology* (Lanham, Md.: University Press of America, 1984), 307. The literature on Christianity's relationship to technology is vast. This volume is a helpful discussion of theological issues and it contains a massive, annotated bibliography of works related to theology and technology.
2. A book scheduled to follow in this Issues of Christian Conscience series will address the issues of medical and biological ethics.
3. Dr. Thomas Wagner, in testimony before the House Judiciary Subcommittee on Courts, Civil Liberties, and the Administration of Justice, June 11, 1987.
4. "Genetic Science for Human Benefit," a policy statement of the National Council of Churches of Christ in the U.S.A., 6. Available from the NCC office, 475 Riverside Drive, New York, NY 10115.

Index

Bosscher, James, 66
Brouwer, Arie, 92, 94
Bultmann, Rudolf, 100
Buridan, 130
Busch, Father Vincent, 79
Byzantium, 129, 133

Cabbala, 136
Calendars, 131
Campus Crusade for Christ,
    15–16
Cancer, 22–23
Carson, Rachel, 73
Chemical industries, 126–27
Chemical waste:
    and effect on health, 72;
    evacuation due to, 72;
    human illness due to, 19–20;
    and Love Canal, 72–73; and
    poisoning, 72; and soil
    contamination, 20; storage
    of, 72. *See also* Waste
Chemicals in food chain, 22, 24
Chernobyl, 23–24, 37–38
China, 128
Chlorofluorocarbons, 22
Christ. *See* Jesus Christ
Christian City (Puerto Rico),
    19–20
Christian College Coalition, 68
Christian Farmers Federations
    (Canada), 77–79; and
    Earthkeeping, 78
Christian Medical College
    (Vellore, India), 80
"Christian Perspectives on
    Stewardship of the Earth's
    Resources" (India), 79–81;
    statement of, 80
Christian Reformed Church(es),
    66, 77–78
Christianity, 131–33, 139,
    152–53, 155; and
    anthropocentrism, 157;
    contrast to paganism, 132;

and culture, 66; and nature,
    13–14, 18; and stewardship,
    68; view of nature, 13–15
Christians, 131, 142; as
    caretakers, 95; as ecologists,
    151; evangelical, 35, 41
Christology, 105, 172–73
Church, 32, 148; advocate of
    social justice, 73; awakening
    of, 26–28; and Council of
    Chalcedon, 105; and
    environment, 49; and
    justice, 41; and politics, 41;
    response of, to ecological
    crisis, 65–82; role of, 32;
    silence of, 73; theology of,
    36–37
Church of the Saviour, 119
Coal, 127; mining of, and acid
    rain, 74; welfare of
    miners, 74
Colonialism, 42–43
Colonists, 44
Commoner, Barry, 39
Communism, 131
Community, 115–16
Composting, 83–85
Concordia Seminary, 124
Constantinople, 44
Contraceptives, 127
Conversion, 41, 97
Copernicus, 129–30, 135
Coral polyp, 124
Cosmos, 62, 169, 173–74
Council of Chalcedon, 105
Covenant, 98–99, 167, 184–85;
    and Abraham, 40; and
    Hosea, 60; and Noah, 40,
    59; with man, 156; with
    nature, 156
Cranfield, C. E. B., 168–69,
    171, 174
Creation, 97–106, 132–33, 146,
    148, 168–69; in agony and
    travail, 171–72; belongs to